Intelligent
Network Video

What key people in the security industry say about *Intelligent Network Video*:

"If you have purchased or are contemplating purchasing this title, you are getting an exceptional, must-read book that was written for every professional that is serious about moving forward in a career with digital surveillance technologies. The author's insights, comprehension and his ability to explain and teach, traverse this exciting field in direct and easily understandable terms. By reading this book, you have taken your first true step toward being part of the future solution."

—Charlie Pierce, President LeapFrog Training & Consulting (LTC) and well-known author of three top-selling CCTV books.

"This book is the most comprehensive resource available to demystify the IP video surveillance market. It is equally as valuable to seasoned professionals as it is to those just entering the field. As a pioneer and leader in the video surveillance market, the author hits all the key issues needed for those interested in doing it right!"

—Sandra Jones, President and Founder, Sandra Jones and Company

"This book is written for those of us who have to "get the job done" using video technology, whether that job involves security planning, system design or operations. Even experienced industry veterans will appreciate the relevance and accuracy of all the material, and should gain an improved ability to discuss the technology with end users and decision makers. This book has definitely improved my thinking about video technology."

—Ray Bernard, PSP, CHS-III, Founder and President, Ray Bernard Consulting Services

"This is the first physical security convergence book to provide the 'Big Picture' on the technical and operational topics of IP networked video surveillance systems, and may well be the most complete and authoritative book written to date."

—Keven Marier, Editor-In-Chief, *IPVS Magazine*

"Opportunities for the use of video abound as never before. As we step back to consider all possible applications for video in surveillance and security, we see clearly that the advancement of video is closely intertwined with advancements and proliferation of other technologies. Fredrik Nilsson's study of the value of networked video shows us a modern problem solved by modern solutions."

—Steve Hunt, CEO and Founder of Hunt Business Intelligence, author of the SecurityDreamer blog

"No question—everyone trying to install new surveillance systems or converge security with IT will benefit from this well-done explanation of why network video is so hot."

—Joe Freeman, CEO, J.P. Freeman Co., Inc. and J.P. Freeman Laboratories, LLC

Intelligent Network Video

Understanding Modern Video Surveillance Systems

Fredrik Nilsson ◆ Axis Communications

CRC Press
Taylor & Francis Group
Boca Raton   London   New York

CRC Press is an imprint of the
Taylor & Francis Group, an **informa** business

CRC Press
Taylor & Francis Group
6000 Broken Sound Parkway NW, Suite 300
Boca Raton, FL 33487-2742

© 2009 by Taylor & Francis Group, LLC
CRC Press is an imprint of Taylor & Francis Group, an Informa business

No claim to original U.S. Government works
Printed in the United States of America on acid-free paper
10 9 8 7 6 5 4 3 2 1

International Standard Book Number-13: 978-1-4200-6156-7 (Softcover)

misprint

Library of Congress Cataloging-in-Publication Data

Nilsson, Fredrik.
    Intelligent network video : understanding modern video surveillance systems / Fredrik Nilsson.
      p. cm.
    "A CRC title."
    Includes bibliographical references and index.
    ISBN 978-1-4200-6156-7 (alk. paper)
    1. Electronic surveillance. 2. Video recording in social change. I. Title.

TK7882.E2N55 2008
621.389'28--dc22
                                    2008013043

Visit the Taylor & Francis Web site at
http://www.taylorandfrancis.com

and the CRC Press Web site at
http://www.crcpress.com

Contents

CONTENTS

CONTENTS

CONTENTS

CONTENTS

CONTENTS

CONTENTS

Introduction

Why is it important to understand network video and intelligent video? Because it represents and drives the most profound change the security industry has seen. We live in a digital and networked world; we use the Web every day, rely on it for banking, share pictures over the Internet, and, increasingly, also use the Internet for voice communication. IP (Internet Protocol) is changing the way we live and work. All markets eventually will be digitized and converge onto the Internet. It is not a matter of *if,* but *when*. The pace of technological development is unstoppable. Moore's law (after Intel co-founder Gordon Moore, who predicted that processing power would double every 18 months) continues to prevail for the foreseeable future as it has for the past 30 years, changing most markets in the process. Now the change has reached the physical security and video surveillance markets. Let us first take a look at three other markets that underwent the same changes.

How the Photography Market Changed

Photography is almost fully digitized but it has taken a long time. The first digital camera was launched in 1994 when the Apple Quick Take 100 stored images digitally. It took another 10 years before the technology was ready for mainstream audiences. In 2002, the first 3- and 4-megapixel cameras, along with ample memory cards, became available at reasonable price points. Today, only a few years later, digital cameras dominate the market with widespread adoption, whereas film cameras have almost disappeared.

The market has converged to a fully digital solution and, with that, the distribution of products also has changed. As IT resellers started to sell cameras, new Web-based services for sharing photos emerged. People started to print their photos on printers at home instead of going to the camera stores or mailing film for processing. Digital photography represents a change in technology, a change in vendors and service providers who address market demands, and a shift in customer behavior. Today, people take ten times more pictures than in the past but develop far fewer. What is next? Soon cameras in cell phones will be good enough to truly replace the traditional camera altogether and the world's largest camera manufacturer might not be Canon, but Nokia. One might wonder if Nokia was listed as a main competitor in the Kodak business plan of 1998. Markets change and manufacturers need to change with them to stay relevant.

How the Music Market Changed

In the mid-1970s, cassette tapes revolutionized the music industry. Consumers had a new medium that was mobile and had the ability to record. The first Sony Walkman came out in the early 1980s and made music easily portable and personal. The next evolution of the Walkman was the CD, a digital technology that provided better audio quality. However, the new technology really did not change the distribution pattern or the end-user experience. CDs are still a physical medium; they require shelf space for storage, need to be bought in person, are cumbersome to copy, and the music is not easily portable.

In 1998, the first MP3 player, Rio Diamond, was launched. By 2002, Apple launched the iPOD and the music industry has never been the same. With MP3 players, the music industry, including distribution, is now fully digitized. Like photography, music has found a fully digital and converged solution. Sony no longer leads the market in manufacturing portable music players; Apple does. Distribution and sales of music continue to change as services such as iTunes take more and more market share. Traditional music stores keep losing market share, and reputable household names such as Tower Records are forced to shut down. Another market has converged to a fully digital solution, and the market dynamics have totally changed.

How the Telecom Market Changed

Another example of a changing market is telecommunications. The telecommunications industry was born when Alexander Graham Bell made the first telephone call in 1876. The industry has seen many gigantic companies emerge, such as Motorola and Ericsson on the vendor side and operators such as AT&T and Telefonica. The early giants were often monopolistic and state controlled. Then in the late 1990s, the first IP telephony solutions began to emerge. Many people frowned at the idea of using the Internet for end-to-end telephony. Common concerns focused on security, latency, and reliability. Many believed it would never work. Today, IP telephony has turned the telecommunications industry upside down. It is not uncommon for consumers to use Comcast as their phone provider and VoIP services such as Skype to dial friends and family overseas. The emergence of telephony companies has caused many large, state-run monopolies to restructure. The number of fixed phone lines in the United States is declining for the first time, and that trend is likely to continue.

More than Just a Change in Products

What is interesting is that any major shift in technology that serves a particular market will have an impact on not only the products themselves but also the route to market, that is, the sales channel and sometimes also the end user. Very rarely is this shift driven by the existing market leader but rather by new market entrants. These entrants — who are digitizing and converging yet another application in our daily lives for the ultimate benefit of the end user — have everything to win and nothing to lose. Convergence and digitization have played a part in the corporate world for a long time. As recently as 15 years ago, most corporate departments had their own computers, operating systems, vendors, and sometimes networks. HR, finance, and production each worked independently. Phone systems were also a separate system. That has completely changed. Functionality and systems have converged onto an open architecture and IP-based platform, saving millions of dollars for large and small companies (Figure I.1).

Change Has Come to the Security Industry

Just as major technological shifts have impacted the music and telecommunications industries and the corporate setting, a major shift is now also happening in the physical security and video surveillance markets. Video surveillance emerged as a viable market about 30 years ago, and the technology has matured year after year. The introduction of the digital video recorder (DVR) in the mid-1990s started to clear a path for a digital solution, but it was really a replacement for the VCR (videocassette recorder)

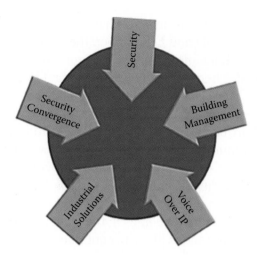

Figure I.1 More and more applications are converging onto IP networks.

and multiplexer. The DVR combined two products into one and added the benefits of digital recording, but it was still a box with analog inputs that simply recorded. New vendors entered the market but buyers did not change, and sales channels remained the same. The DVR signified an evolution in technology — not a revolution.

After a few years, DVRs were equipped with a communication interface that enabled the retrieval of recorded or live video from a remote location. Initially, the interface was a standard serial port that connected to a phone modem, but eventually it became an Ethernet port, requiring the first collaboration with the IT department. However, for the most part, security systems within an organization operated under the radar of CEOs and CIOs.

In 1996, the world's first network camera saw the dawn of light. That camera was the AXIS 200 (Figure I.2), a camera that could send a small-resolution image at one frame per second, and four frames per minute if one wanted full resolution. The AXIS 200 was a product that was poised to replace and overtake the Web camera market for remote monitoring and Web attraction applications. For security applications, however, that camera was far from sufficient. Fast-forward a few years and the network camera had evolved significantly. It could deliver 30 frames per second, had built-in video motion detection, and had an image quality that was similar to analog cameras. All of a sudden, cameras were deployed for security purposes.

Back in 1996, most networks were 10-Mbit networks, and just the thought of putting surveillance video on those networks was unthinkable. By the year 2000, 100-Mbit networks were commonplace, and by 2007, a common enterprise-class, 48-port network switch enabled Gigabit performance on every port and had the ability to stream video from thousands

Figure I.2 AXIS 200, the world's first network camera.

of network cameras at full frame rate via the 10-Gigabit backplane. Remember that Moore's law is still prevailing and technology development happens very fast — yesterday's bottlenecks might become tomorrow's opportunities.

Because of the open interface in the IT industry, independent software companies that build applications for video management started to emerge in Germany, the United States, Spain, Denmark, Japan, and Canada — essentially all over the globe. A new market was starting to take shape: the network video market.

The Changing Face of the End User

When the attacks of September 11th happened, security rose to the top of the agenda in every state, federal, and Fortune 1000 organization— not only physical security but also IT security. No longer was security just an issue of amateur hackers wanting to prove themselves by hacking Web pages of large companies. Now, terrorists wanted to disrupt financial systems and our daily lives. The post-9/11 world called for more integration of different departments and higher standards for the systems implemented. One is starting to see a transition to a structure where the security manager reports to the Chief Security Officer responsible for both physical and IT security. The purpose is twofold: (1) to integrate and coordinate all security, and (2) to migrate the security systems over to an open architecture and IP-based platform (Figure I.3).

What Is Open All About?

In IT, there is a lot of discussion about open systems, open architectures, open standards, and open APIs (application programming interfaces). For a systems integrator or an end user, those are key words. Openness, simply

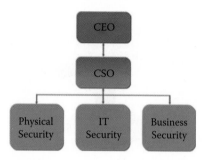

Figure I.3 With the increased importance of security, the Chief Security Officer (CSO) position has been put in place, managing both physical and IT security.

put, will provide a better system at a lower cost. Evolution in the IT industry has shown that this is really the only way to conduct a successful, long-term business. Look at various vendors that are focusing on specific disciplines: EMC for storage, Cisco for networks, Microsoft for operating systems, and Intel for processors. To be really successful, a company needs to focus on one discipline and identify partnerships or work in ecosystems to provide the rest of the pieces that make a complete system for the end user (Figure I.4). This also is why the systems integrator is a key player — to ensure that the pieces are properly integrated or "glued together." To a large extent, the glue that puts it all together is the Internet Protocol (IP). Compatibility and exchangeability are ensured by using open standards, and several standardization bodies, such as the IEEE (Institute of Electrical and Electronics

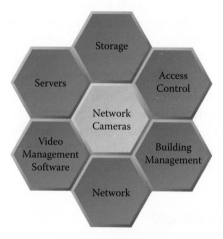

Figure I.4 A best-of-breed network video system consists of many components from different vendors, all communicating and interoperating via open interfaces.

Engineers), IETF (Internet Engineering Task Force), and MPEG (Moving Picture Experts Group), help make that happen.

History has shown that focusing on one discipline and partnering with other companies for other components were and continue to be the winning strategy in the fast-moving IT world. The beneficiary is the end user, who gets ever-increasing functionality at a lower cost, year after year. The alternative strategy would be a proprietary solution, with all components from one vendor. A proprietary system was the strategy in the early days of IT — back in the 1970s and 1980s — and was executed by companies such as Digital Equipment and Wang. Initially, this strategy was very successful, but when IP networks and standards emerged, it became very difficult for these companies to keep up. They lost market leadership, and some even went out of business.

How Open Is Open?

Proprietary systems can be great for a vendor because it means that the end user does not have to look anywhere else for system components. In a proprietary system, all system components are procured from one vendor. This kind of system is like a monopoly because prices — normally higher prices — are dictated by the vendor. Monopolies also typically mean longer product generation cycles. However, monopolies usually are not associated with an open and successful market. The trick is knowing what *open* really means.

Being *open* is a buzzword nowadays because many companies have shown that this is the most successful way to conduct long-term business. Today, it would be difficult to find a company that would market a system as being proprietary. However, it is more advantageous for a leading company that owns most of the market share to lock in a customer and make the system a little bit less open in an effort to control the customer's buying habits. The end user must be aware that there are many different degrees of openness. Watch out for companies that manufacture all of the components in a certain system, do not have openly published APIs, have a limited number of partnerships, or do not 100 percent comply with open standards. One will likely end up being locked in, with higher system costs as a result.

Why Intelligent Video Will Become Essential

One of the benefits of an IP-based video surveillance system is the ability to scale. Although a 50-camera system was considered large back in the early 1990s, it is small compared to what an IP-based system enables. For example, an IP-based system can accommodate thousands or even tens of thousands of cameras in one integrated system. Who will monitor all the video? Research shows that even after 20 minutes, an operator will miss important events using a quad monitor. Already today, there are more than two million analog cameras in London alone, and 15 million in the United States. This is where video intelligence or video content analysis comes into play.

Video management systems, and even cameras, can do the monitoring and alert the operator to certain types of events, such as when a camera has been tampered with, if someone has entered the track in a subway, or in applications where there is a need to count the number of people going into a certain area. The intelligent video market is still an early market, but it will be an important driver for further deployment and usage of video systems. Intelligent video systems will be truly scalable and useful if the intelligence resides "at the edge" of the system, that is, in the network cameras.

Convergence: Here to Stay and Here to Change Our Lives

According to *Webster's Dictionary*, convergence means "the merging of distinct technologies, industries, or devices into a unified whole." Convergence changes many things: there is convergence of the technology, the route to the market, and the roles within the end-user organizations. Security is one of the last functions within many organizations that has not yet converged onto standard IT equipment. HR and finance departments did it a decade ago. It did not mean that those departments lost any control of their applications; on the contrary, it demonstrated the need to work more closely with the IT department to let it know what their needs were. It also meant that once those needs were clarified, the operation and maintenance of those systems were in the capable hands of the IT department, and there was one less worry for the HR and finance departments to think about and more time could be spent on their professions.

For many years, the discussion was whether convergence would happen in the security industry, and many people had their doubts. Since 2001, security systems have converged onto IP-based networks and the change is now widely accepted. 2006 and 2007 saw the beginning of the hypergrowth phase of network video, with the installation of thousands of cameras in the education, retail, and transportation markets. The writing is on the wall, and the question emerging is *"How fast?"*. How fast will security solutions within all organizations evolve to be yet another application on the IP network? The technology is here, and it is here to stay, so the only alternative is to embrace it.

About This Book

Open systems and more choices are advantageous but they also call for a more educated end user and systems integrator. That is the main purpose of this book: to educate anyone interested in jumping onto the fast-moving intelligent network video train and benefit from the possibilities.

This book takes the reader through a logical tour of the building blocks of intelligent network video, including network cameras, video encoders, networks, storage, servers, and video management. Intelligent video, system design, and the cost of network video are discussed in detail at the end of the book. There are also a few chapters with fairly technical information. Such chapter titles end with the word "Technologies," indicating that they are suitable for the more technical reader. The book also comes with a system design tool DVD, a tool that helps one design a video surveillance project based on resolution, compression, frame rate, bandwidth, and storage considerations.

As one navigates through the fascinating world of network video, the authors hope the book will serve as a guide in helping implement successful IP-based video surveillance systems.

About the Author

Fredrik Nilsson currently serves as the general manager for Axis Communications, Inc., overseeing the company's operations in North America. He has been with Axis for more than 10 years and has been general manager of the North American region since 2003. In that time, he has helped the company increase revenue fivefold and has been instrumental in leading the industry shift from analog closed circuit television to network video.

Fredrik Nilsson also serves on the SIA (Security Industry Association) Board of Directors. He is a trusted industry speaker and has spoken at more than 20 conferences, including providing the keynote address at TechSec Solutions, the premier conference on IP-ready security technology, and at influential shows such as Securing New Ground, ASIS Emerging Trends in Security, ISC West, ISC East, and Interop.

He has been in top publications such as *The New York Times*, *USA Today*, and *The Washington Post*. He has also appeared on television shows such as CNN Headline News, CNBC's "Wake Up Call," and the Fox News Channel's "Fox & Friends." He has written articles for and been quoted in many publications in the security industry, including *Access Control & Security Systems, Security Products & Technology, Security*

Management, and *Security Systems News.* He is also a regular columnist for SecurityInfoWatch.com.

Prior to working for Axis, he served as a product manager for ABB, a global leader in power and automation technologies. A graduate of the Lund Institute of Technology, Fredrik Nilsson holds a master's degree in electrical engineering, which he followed up with post-graduate studies in economics.

Acknowledgments

Writing a book, like any large project you take on in life, cannot be completed without tremendous support from colleagues, friends and family. This book was no different.

I am fortunate enough to work for Axis Communications, a company that saw the value in the book, and not only gave me the time, but more importantly, all the required internal resources to complete it. All of you who helped, from proofreading to writing whole chapters, are aware of your invaluable contributions and my sincere thanks go out to all of you.

I am also lucky to be part of an industry with some outstanding people who are in the industry not only to make a living but also to help provide a safer and more secure life for all of us. Since moving to the United States six years ago, I have gained tremendous knowledge from many generous people willing to teach me from their lifelong experiences, many whom I call friends.

This book would not have existed without the understanding and support from my wife. Thanks for letting me spend so many weekends and nights on this project. I love you.

Chapter 1

The Evolution of Video Surveillance Systems

Video surveillance, more commonly called CCTV (closed-circuit television), is an industry that is more than 30 years old and one that has had its share of technology changes. As in any other industry, end users' ever-increasing demands on the products and solutions are driving the changes, and evolving technologies are helping to support them. In the video surveillance market, the demands include:

- Better image quality
- Simplified installation and maintenance
- More secure and reliable technology
- Longer retention of recorded video
- Reduction in costs
- Size and scalability
- Remote monitoring capabilities
- Integration with other systems
- More built-in system intelligence

To meet these requirements, video surveillance has experienced a number of technology shifts. The latest is the shift from analog CCTV surveillance to fully digital, network-based video surveillance systems.

Video surveillance systems started out as 100-percent analog systems and are gradually becoming digital. Today's systems, using network cameras and PC (personal computer) servers for video recording in a fully digital system, have come a long way from the early analog tube cameras, which were connected to a VCR (videocassette recorder).

In between the fully analog and fully digital systems, there are several solutions that are partly digital, that is, systems that incorporate both digital and analog devices. This has led to some confusion in the video surveillance industry today, as some talk about a "digital" system to mean analog cameras that connect to a DVR (digital video recorder), whereas others use the term to describe a network video system with network cameras. Although there are digital components in both systems, there are some very important distinctions to make between each of the systems.

The sections below outline the evolution of video surveillance systems. Different system configurations, from fully analog to fully digital, are explained, along with the benefits of each configuration. The systems described in Sections 1.2 and 1.3 constitute partly "digital" video systems. Only the systems described in Sections 1.4 and 1.5 are true network video systems in which video streams are continuously being transported over an IP network, providing full scalability and flexibility.

1.1 VCR-Based Analog CCTV Systems

The traditional analog CCTV system involved the use of analog cameras that were connected to a VCR for recording video (Figure 1.1). The system was completely analog. The VCR used the same type of cassettes as those sold for a home VCR. Each camera needed its own coax cable to run from the camera all the way to the VCR. The video was not compressed, and when recording at full frame rate, one tape lasted a maximum of eight hours. Eventually, a so-called time lapse mode was incorporated into the VCRs to make the tape last longer. The time lapse mode enabled the recording of every second, fourth, eighth, or sixteenth image. That

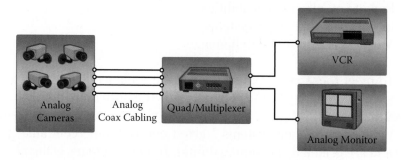

Figure 1.1 Diagram showing a classical analog video surveillance system.

was how the video surveillance industry came up with such specifications as 15 fps (frames per second), 7.5 fps, 3.75 fps, and 1.875 fps, because these were the only recording frame rates possible in analog systems that used time lapse recording. If several cameras were used, quads became another important system component. A quad simply took inputs from four cameras and created one video signal output to show four different images on one screen; hence, the name "quad." This invention made the system a bit more scalable but at the expense of lower resolution.

In even larger systems, multiplexers became commonplace. A multiplexer combined the video signals from several cameras into a multiplexed video signal. This made it possible to record even more cameras, often 16 on one device. The multiplexer also made it possible to map selected cameras to specific viewing monitors in a control room. Still, all equipment and all signals were analog. To monitor the video, analog monitors connected to a VCR, quad, or multiplexer.

Although analog systems functioned well, the drawbacks included limitations in scalability and the need to maintain VCRs and manually change tapes. In addition, the quality of the recordings deteriorated over time. The cameras, for a long time, were also black and white. Today, most analog cameras are in color.

1.2 DVR-Based Analog CCTV Systems

By the mid-1990s, the video surveillance industry saw its first digital revolution with the introduction of the DVR. The DVR, with its hard drives, replaced the VCR as the recording medium (Figure 1.2). The video was digitized and then compressed to store as many days' worth of video as possible.

With early DVRs, hard disk space was limited, so the recording duration was limited or a lower frame rate had to be used. Due to the

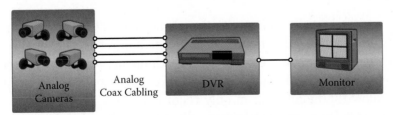

Figure 1.2 A surveillance system with analog cameras connected to a DVR, which includes the quad or multiplexer functionality and provides digital recording.

limitations in hard disk space, many manufacturers developed proprietary compression algorithms. Although they might have worked well, end users were tied to one manufacturer's tools when it came to replaying the video. As the cost of hard disk space decreased dramatically over the years and standard compression algorithms such as MPEG-4 became available and widely accepted, most manufacturers gave up their proprietary compression in favor of standards — to the benefit of end users.

Most DVRs had several video inputs, typically 4, 16, or 32, which meant they also included the functionality of the quad or multiplexer. Hence, DVRs replaced the multiplexer as well as the VCR and thereby reduced the number of components in the CCTV system.

The introduction of the DVR system provided the following major advantages:

- No tapes and tape changes
- Consistent recording quality
- Ability to quickly search through recorded video

Early DVRs used analog monitors such as TV sets for showing video. However, because the DVR made digital video available, it became possible to network and transmit the digital video over longer distances. This function was first addressed by connecting a phone modem to a serial port on the DVR. Later, the phone modem was built into the DVR itself. Although the ability to monitor the video remotely via a PC was a great benefit, the actual functionality was not extremely useful because the bandwidth available with phone modems was too low, often in the 10- to 50-kbps range. That meant very low frame rates, low resolution, or highly compressed video, which made the video more or less useless.

1.3 Network DVR-Based Analog CCTV Systems

DVRs were eventually equipped with an Ethernet port for network connectivity. This introduced network DVRs to the market and enabled remote video monitoring using PCs (Figure 1.3). Some network DVR systems in use today enable the monitoring of both live and recorded video, whereas some allow the monitoring of only recorded video. Furthermore, some systems require a special Windows client to monitor the video, whereas others use a standard Web browser; the latter makes remote monitoring more flexible.

The network DVR system provides the following advantages:

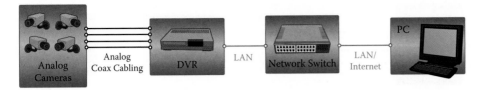

Figure 1.3 A system that shows how analog cameras can be networked using a network DVR for remote monitoring of live and recorded video.

- Remote monitoring of video via a PC
- Remote operation of the system

Although DVRs provided great improvements over VCRs, they also had some inherent downsides. The DVR was burdened with many tasks such as the digitization of video from all cameras, video compression, recording, and networking. Additionally, it was a "black box" solution, that is, proprietary hardware with preloaded software, which often forced the end user to source spare parts from one manufacturer, making maintenance and upgrading expensive. Virus protection was also difficult to implement. Although the DVR was often a Windows-based machine, the proprietary interface did not allow for virus protection. In addition, the DVR offered limited scalability. Most DVRs offered 16 or 32 inputs, which made it difficult to cost-effectively build systems that were not multiples of 16, for example, systems with 10 or 35 cameras.

1.4 Video Encoder-Based Network Video Systems

The first step into a networked video system based on an open platform came with the introduction of the video encoder, which is also often called a video server.

A video encoder connects to analog cameras and digitizes and compresses the video. It then sends the video over an IP network via a network switch to a PC server that runs video management software for monitoring and recording (Figure 1.4). This is a true network video system because the video is consistently sent over an IP network. In essence, the tasks previously performed by the DVR are now divided up — with the digitization and compression being done by the video encoder and the recording by the PC server — thus providing better scalability.

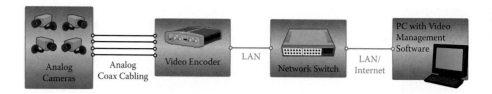

Figure 1.4 A true network video system, where video is continuously transported over an IP network. It uses a video encoder as the cornerstone to migrate the analog security system into an open IP-based video solution.

A video encoder–based network video system has the following advantages:

- Use of standard network and PC server hardware for video recording and management
- Scalability in steps of one camera at a time
- Possibility to record off site
- Future-proof because the system is expanded easily by incorporating network cameras

1.4.1 NVRs and Hybrid DVRs

Alternatives to the open platform (based on a PC with video management software installed) are also possible with the availability of different types of NVRs (network video recorders; Figure 1.5) and hybrid DVRs. An NVR or hybrid DVR is a proprietary hardware box with preinstalled video management software for managing video from video encoders or network cameras. The NVR handles only network video inputs, whereas

Figure 1.5 The NVR is a hardware box with preinstalled video management software that makes installation simpler but lacks the flexibility of an open-platform system based on a standard PC server.

the hybrid DVR can handle both network video as well as analog inputs in parallel. The benefit of using an NVR or hybrid DVR is the ease of installation because the recording and video management functionalities are made available all in one box — similar to a DVR. The NVR or hybrid DVR solution is popular in smaller systems with 4 to 16 cameras. However, it also maintains some of the drawbacks of the traditional DVR. NVRs and hybrid DVRs use a proprietary platform that is more expensive to purchase, maintain, and upgrade, and they are often difficult to maintain on a corporate IT network.

1.5 Network Camera-Based Network Video Systems

A network camera, also commonly called an IP camera, is, as its name describes, a camera with an IP network connection. In a network camera–based network video system, video is transported over an IP network via network switches and is recorded to a PC server with video management software installed (Figure 1.6). This represents a true network video system. The system is fully digital because no analog components are used.

One of the greatest benefits of a network camera is that once images are captured, they are digitized once inside the camera and remain digital throughout the system, which provides for high and consistent image quality. This is not the case with analog cameras. Although most analog cameras today are called "digital," they have an analog output, and this can lead to some confusion. Analog cameras do digitize captured images to provide image-enhancing functions. However, these images are then converted back to analog video. It is important to know that with every conversion from analog to digital, or from digital to analog, there is some

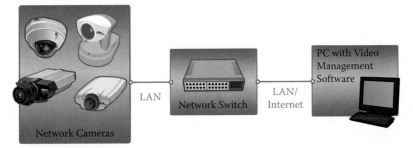

Figure 1.6 This diagram shows a true network video system where videos from network cameras are continuously transported over an IP network. This system takes full advantage of digital technology and provides consistent image quality from the cameras to the viewer at any location.

loss of video quality. Additionally, analog signals degrade when transported over long cables and over time if stored on tape. Therefore, video ideally should be digitized once and stay digital throughout the system.

The benefit of using an IP-based network is that one can use the network for more than just transporting video. IP networks provide a means for several network cameras to share the same physical cable. In addition, the network can carry power to network cameras and information to and from the outputs and input contacts of the cameras. It can also carry two-way audio, as well as pan, tilt, and zoom commands if a camera has that functionality. Furthermore, an IP network enables network cameras to be configured remotely and allows video and other data that are sent over the network to reach virtually any location with no degradation in quality. All in all, the network provides an extremely flexible and cost-efficient medium for all communications within a network video surveillance system. The scalability of network video provides opportunities to build video surveillance systems with hundreds or even thousands of cameras.

A network camera–based network video system provides the following advantages:

- Ability to use high-resolution (megapixel) cameras
- Consistent image quality, regardless of distance
- Ability to use Power over Ethernet and wireless functionality
- Full access to functionalities such as pan, tilt, and zoom; audio and digital inputs and outputs over IP, together with video
- Camera settings and system adjustments over IP
- Full flexibility and scalability

Although a network camera can be compared to an analog camera attached to a video encoder, a network camera can offer many more functionalities that extend beyond the capabilities of a system involving an analog camera and a video encoder. Because a network camera has built-in computing power, it opens up the possibility for video intelligence, also called video analytics, at the edge of the system, that is, within the camera. This is expected to be the next big trend within video surveillance because there is a necessity to manage and analyze video effectively, especially in large systems.

The network camera is a key driver in the network video revolution. Network cameras have fully caught up with analog camera technology and now meet the same requirements and specifications. Moreover, network cameras surpass analog camera performance in many important areas, such as image quality, resolution, and built-in intelligence.

Chapter 2

The Components of Network Video

To understand the scope and potential of an integrated, fully digitized system, let us first examine the core components of a network video system: the network camera, the video encoder (also known as the video server), the network, the server and storage, and the video management software.

Although all components are necessary, it is interesting to note that the network, the server, and the storage components are all standard IT equipment, and the ability to use commercial-off-the-shelf (COTS) equipment is one of the main benefits of network video. The other three components — the network camera, the video encoder, and the video management software — are unique to network video and are the cornerstones of network video solutions (see Figure 2.1). Another component that is emerging is intelligent video, or video analytics, which can reside in the network camera, video encoder, or video management software.

2.1 Where Is Network Video Used?

Network video, often also called IP-based video surveillance or IP-Surveillance, is a system that gives users the ability to monitor and record video, as well as audio, over an IP network (LAN, WAN, or Internet).

Unlike analog video systems, network video uses an IP-based network rather than dedicated point-to-point cabling as the backbone for transporting video and audio. In a network video application, digitized video

Figure 2.1 A network video system consists of many different components, including network cameras, video encoders, and video management software. The other components including the network, storage, and servers are all standard IT equipment.

and audio streams are sent over wired or wireless IP networks, enabling video monitoring and recording from anywhere on the network.

Network video can be used in an almost-unlimited number of applications; however, most of its uses fall into security surveillance or remote monitoring:

- *Security surveillance.* Network video's advanced functionalities are highly suited to applications involved in security surveillance. The flexibility of digital technology enhances security personnel's ability to protect people, property, and assets. Such systems are therefore an especially attractive option for companies currently using CCTV (closed-circuit television).
- *Remote monitoring.* Network video gives users the ability to gather information at all key points of an operation and view it in real-time. This makes the technology ideal for monitoring equipment, people, and places — both locally and remotely. Application examples include traffic and production line monitoring, and the monitoring of multiple retail locations.

As is the case with most new technologies, it is in a few key verticals where network video has been first deployed and its benefits realized. The main vertical markets where network video systems have been successfully installed include:

- *Education: security and remote monitoring of school playgrounds, hallways, and classrooms.* Many schools today lack surveillance systems but have a very strong IT infrastructure in place for other applications such as data and voice. In such situations where no surveillance installation exists, a network video solution presents a very favorable alternative to an analog system from a cost perspective, because new cabling often is not needed.

- *Transportation: video surveillance of railway stations and tracks, parking lots and garages, highways, and airports.* Many of today's post-9/11 installations in the transportation sector are high-profile installations where only the best systems — involving high image quality, high frame rates, and long retention times — are deployed. Such features are only achievable in a network video system. Lately, several deployments also have been made in the mobile transportation environment, such as buses, trains, and cruise ships. This development is driven by the improved video quality in network cameras that use progressive scan technology.

- *Banking: traditional security applications in bank branches, headquarters, and ATM locations.* Banks have been using surveillance for a long time, and although most installations are still analog, network-based video is starting to make inroads, especially in banks that value high image quality and want the ability to easily identify people in surveillance video.

- *Government: for surveillance purposes to provide safe and secure public environments.* Network video systems are installed in, for example, state facilities, air force bases, courthouses, and prisons. The reasons include secure communication, scalability, and image quality. Surveillance of city centers is also becoming an important vertical for network video, especially in combination with different types of wireless networks.

- *Retail: for security and remote monitoring purposes to make store management easier and more efficient.* Retail is the biggest vertical market for video surveillance. Many large retail chains have 50 to 300 cameras deployed per store, helping their Loss Prevention or Assets Protections team do a better job and reduce shrinkage. This is a price-sensitive market, and the fact that network video has started to penetrate the retail segment shows that the cost of network video is now lower than analog systems in large installations. Additionally, network video makes integration with point-of-sales systems easier.

- *Industrial: monitoring manufacturing processes, logistic systems, and warehouse and stock control systems.* Industrial environments have been using Ethernet and TCP/IP for a long time, making network video a natural fit for monitoring and increasing efficiencies.

2.2 The Network Camera

A network camera, often also called an IP camera, can be described as a camera and computer combined into one unit. It captures and sends live images directly over an IP network, enabling authorized users to locally or remotely view, store, and manage video over a standard IP-based network infrastructure. See Figures 2.2 and 2.3.

A network camera has its own IP address. It is connected to a network and has a built-in Web server, FTP server, FTP client, e-mail client, alarm management, programmability, and much more. A network camera operates as an independent server on a network and can be placed wherever there is an IP network connection. A Web camera, on the other hand, is something totally different. A Web camera only operates when it is connected to a PC via a USB (Universal Serial Bus) or IEEE 1394 port; and to use it, software must be installed on the PC.

In addition to video, a network camera also can support other functionalities such as audio, alarm activation via digital inputs and outputs, and serial communications.

2.2.1 Comparing a Network Camera and an Analog Camera

In recent years, network camera technology has caught up with the analog camera and now meets the same requirements and specifications.

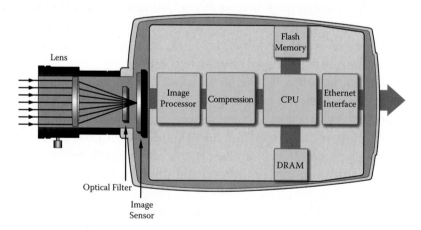

Figure 2.2 Inside a network camera. The image processor and compression is handled by an ASIC (application-specific integrated circuit) or a DSP (digital signal processor). The CPU (central processing unit), flash memory, and DRAM (dynamic random access memory) are specialized for network applications.

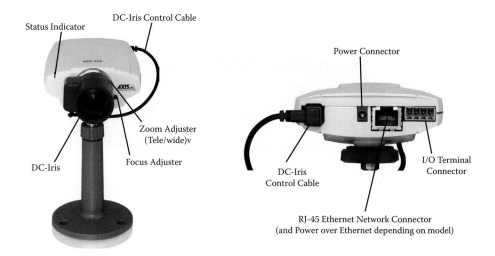

Figure 2.3 Picture of a typical network camera with some important features pointed out (at left, front view; at right, rear view).

Network cameras even surpass the performance of analog cameras by offering a number of advanced features, such as higher image quality, higher resolutions, and built-in intelligence.

An analog camera is a one-directional signal carrier that ends at the recording device, whereas a network camera is fully bi-directional allowing information to be sent and received, and can be an integrated part of a larger, scalable system. A network camera communicates with several applications in parallel to perform various tasks, such as detecting motion or sending different streams of video.

Read more about network cameras in Chapters 3 and 4. For more detailed information about the technology inside network cameras, including such topics as compression and audio, refer to Chapters 5 and 6.

2.3 The Video Encoder

A video encoder makes it possible to integrate an analog camera into a network video system without having to discard existing analog equipment. It brings new functionalities to analog equipment and eliminates the need for dedicated equipment such as coaxial cabling, analog monitors, and DVRs — the latter becoming unnecessary as video recording can be done using standard PC servers.

A video encoder typically has between one and four analog ports for analog cameras to plug into, as well as an Ethernet port for connection to

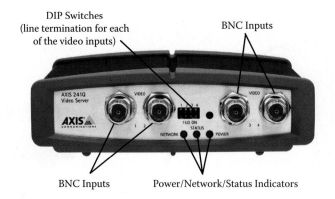

DIP Switches
(line termination for each
of the video inputs)

BNC Inputs

BNC Inputs

Power/Network/Status Indicators

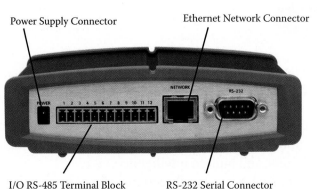

Power Supply Connector

Ethernet Network Connector

I/O RS-485 Terminal Block

RS-232 Serial Connector

Figure 2.4 Picture of a typical video encoder with some important features pointed out (on top, front view; on bottom, rear view).

a network (Figure 2.4). Like network cameras, a video encoder contains a built-in Web server, a compression chip, and an operating system so that incoming analog feeds can be converted into digital video, sent, and recorded over a computer network for easier accessibility and viewing.

In addition to the video input, a video encoder can also support other functionalities such as audio, alarm activation via digital inputs and outputs, and the control of PTZ (pan, tilt, zoom) mechanisms through serial ports. A video encoder also can be connected to a wide variety of specialized cameras, such as a highly sensitive thermal camera, a miniature camera, or a microscope camera.

For large installations with existing analog cameras, rack-mounted versions of video encoders are available and can offer as many as 84 ports in one rack solution (Figure 2.5).

With millions of analog cameras in operation, video encoders will continue to play an important role in video surveillance systems. Chapter 7 provides more information about video encoders.

Figure 2.5 For large installations with coax cabling already installed, a video encoder rack is a popular solution.

2.4 The Network

One of the most obvious benefits of network video is the ability to use standard wired or wireless IP-based networks. Not only are IP-based networks cost efficient to deploy, but in many cases they already exist and can be used also to carry power to a network device such as a network camera. IP networks exist in many different forms, from small wireless local area networks (LANs) in homes to corporate networks, citywide wireless deployments, and the Internet. IP-based networks are used by almost everyone on a daily basis, whether at home or at work. Most buildings today come wired for IP networks.

The growing use of IP networks means that companies are spending an enormous amount of money on research and development to constantly improve the functionality, speed, and security of such networks. Whereas bandwidth was a bottleneck only a few years ago, today's gigabit networks provide more than enough bandwidth for content-rich applications such as network video. Even mobile cellular networks are beginning to have sufficient bandwidth for video distribution at reasonable costs.

As everyone relies on access to a network in their daily lives, network security is becoming an increasingly important issue, especially in sensitive applications such as bank transactions, government networks, and now also video surveillance. It is important, therefore, to understand network security technologies such as 802.1X, HTTPS, and firewalls.

Chapters 8 and 9 provide more information about wired and wireless networks, and Chapter 10 provides more technical details about network technologies and includes topics such as network security.

2.5 The Server and Storage

The server and storage used in network video are components based on standard equipment from the IT industry. This means that a network video system can benefit from rapid innovation in processor and storage technologies. Being a growing, multi-billion-dollar market means heavy investments in research and development can produce servers with twice the performance and storage space every 18 months at the same cost.

Most servers that run video management applications are PC servers with a Windows operating system. Depending on the performance required, a dual- or even quad-core processor can be used, making it possible to manage video from 50 cameras at full frame rate, or more than 100 cameras at lower frame rates. Enterprise-class servers usually deploy higher-end SCSI hard drives, whereas lower-cost ATA drives can be appropriate in some smaller applications. To ensure redundancy, RAID (Redundant Array of Independent Disks) arrays are often deployed. For smaller installations of up to 25 cameras, a regular server with a single-core processor would be appropriate. In larger network video installations, network attached storage (NAS) or storage area networks (SANs) are deployed.

Today, virtually any size storage system can be built, so knowing what technologies are available and what level of performance and redundancy can be provided, and at what costs, is important in successfully installing an network video system. For more information about servers and storage, see Chapter 11.

2.6 Video Management Software

For very small systems, particularly where only one or a few cameras are viewed at the same time, a standard Web browser that utilizes the Web interface built into a network camera or video encoder provides adequate video management functions. However, to view several cameras at the same time requires dedicated video management software.

Video management software supplies the basis for video management, monitoring, analysis, and recording (Figure 2.6). Today there are hundreds of different video management applications available from different companies around the world, covering different operating systems (Windows, UNIX, Linux, and Mac OS), vertical markets, languages, scalability requirements, and integration possibilities with other systems such as building management, access control, and industrial control. Open plat-

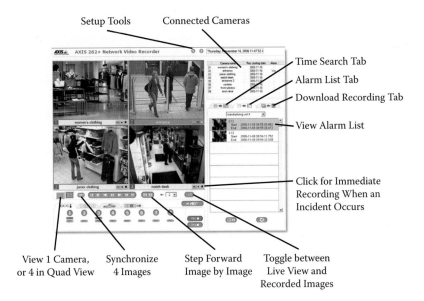

Figure 2.6 Screenshot of video management software showing the interface the operator would use.

form solutions run on "off-the-shelf" hardware, with components selected for maximum performance.

Manufacturers of network cameras and video encoders are publishing application programming interfaces (APIs) to ensure integration and compatibility with video management software. The more openly the API is published, preferably on the Web and free of charge, the better the selection of video management software and the tighter the integration is for a given network camera or video encoder.

Some of the common features provided by most video management software are:

- *Simultaneous viewing and recording of live video from multiple cameras:* video management enables multiple users to view several different cameras at the same time and allow recordings to take place simultaneously.
- *Several recording modes:* continuous, manual, scheduled, on alarm, and on motion detection. Video motion detection defines activity by analyzing data and differences in a series of images. Video motion detection can be performed by the camera, which is preferable, or by the software.
- *Multiple search functions for recorded events:* areas of interest can be defined and video from several cameras can be replayed at the same time.

- *Camera management:* video management systems allow users to administer and manage cameras from a single interface. This is useful for tasks such as detecting cameras on the network; managing IP addresses; and setting resolution, compression, and security levels.
- *Remote access:* via a Web browser, client software, and even PDA client.
- *Control of PTZ and dome cameras:* control can be via a joystick or mouse that is controlled by an operator, or it can be done automatically via guard tours controlled by the software or set in the cameras.
- *Configuration of I/Os:* enables video to be sent and recorded, and alarms to be sent in response to external sensors. This allows remote monitoring stations to become immediately aware of a change in a monitored environment.
- *Alarm management:* the software can sound an alarm, display pop up windows, send e-mails, or send text messages (Short Message Service, SMS) to cell phones.
- *Audio support:* real-time audio support, for either live or recorded audio.

Video management systems are easily scalable because cameras can be added one at a time, and some systems can scale to thousands of cameras. Open systems are suitable for scenarios where large numbers of cameras are deployed. They also make it easier to add functionalities to the system, such as increased or external storage, firewalls, virus protection, and intelligent video algorithms to the system. Some video management systems use a Web interface to access the video from any type of computer platform. Web interfaces allow video online management from anywhere in the world, using the proper safeguards (e.g., password protection and IP address filtering).

Video management systems based on open platforms have another advantage in that they can be more easily integrated with other systems such as access control, building management, and industrial control. This allows users to manage video and other building controls through a single program and interface. Integrating a video surveillance system with an access control system, for example, allows for capturing video at all entrance and exit points, enabling photos in a badge system to be matched against images of the person actually using the access card.

Chapter 12 provides additional information on video management software.

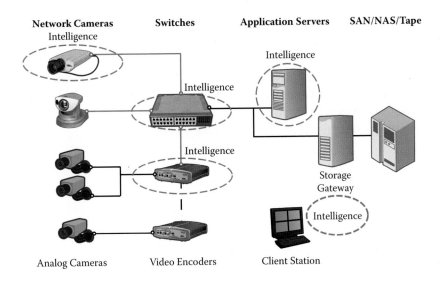

Figure 2.7 Video intelligence is a requirement in systems where a fast response time is required or a very large number of cameras are to be managed proactively. For scalability reasons, intelligence should preferably reside "at the edge," that is, in the network cameras or video encoders, although it also is possible for a server or a client running video management software to deploy it.

2.7 Intelligent Video

Intelligent video (Figure 2.7), also called video intelligence, video content analysis, or video analytics, is a quickly emerging market driven by a desire to get more out of a video surveillance system — whether it is obtaining faster response times, deploying fewer operators per camera, or extracting information for other purposes than surveillance, such as people counting. Research from Sandia National Laboratories in the United States has shown that an operator is likely to miss important information after only 20 minutes in front of a monitor; intelligent video can help operators cover more cameras and respond more quickly.

The architecture of an intelligent video system can be either centralized or distributed. In a centralized system, all intelligence resides in the recording server. The main benefit of a centralized system is that the server, in most cases, is an open platform, so adding functionality is simple. Intelligent video algorithms do consume a lot of computing power, which means that only a few cameras can be managed at any given time by one server. In a distributed system, the intelligence is distributed to the edge and resides in the network camera or video encoder. The main benefit of this is that the analysis is done locally in the camera, making the system fully scalable and potentially enabling reduced bandwidth usage because the camera or video encoder can determine whether to send video to the server based on the video content. Intelligent video is mostly used in high-risk transportation, government, and retail applications today. The functionality includes camera tampering alarm, people counting, license plate recognition, and object tracking. For more information on intelligent video, see Chapters 13 and 14.

Chapter 3

Network Cameras

A wide variety of network cameras, also commonly called IP cameras, is available to meet most requirements and system sizes. Today, network cameras offer more benefits than analog cameras, including better image quality, higher resolution, and built-in intelligence.

This chapter discusses the components of a network camera and the different camera types. It also provides in-depth information about network dome and PTZ (pan, tilt, zoom) cameras, non-mechanical-PTZ cameras, and megapixel network cameras. Additional information about camera technologies is provided in Chapter 4.

3.1 Network Camera Components

A network camera is, as its name describes, a camera with a network connection. The main components of a network camera include (see Figure 3.1):

- A lens for focusing the image on the image sensor
- An image sensor, either CCD (charge-coupled device) or CMOS (complementary metal oxide semiconductor)
- One or several processors for image processing, compression, video analysis, and networking functionality
- Memory for storing the firmware code of the network camera (using Flash) as well as local recording of video clips and events (using RAM)

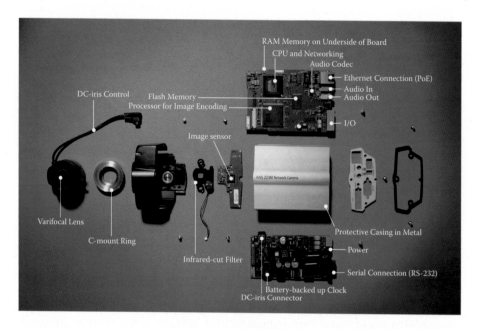

Figure 3.1 Components of a network camera.

In the analog world, cameras exist primarily in two formats: (1) PAL (the European CCTV standard) or (2) NTSC (the American CCTV standard). In the network video world, those standards are not relevant because network cameras connect to Ethernet cables that transport Motion JPEG and MPEG-4 compressed video, which are worldwide standards — a benefit when planning and servicing worldwide deployments.

A network camera can incorporate many advanced functionalities, for example, video motion detection, alarm handling, audio monitoring, and audio alarm. The networking functionality is also important and should include all the latest security and IP protocols. Different cameras come with different performance; some provide less than full frame rate (30 fps) of video, whereas others can provide 60 fps or several video streams at full frame rate simultaneously.

3.2 Types of Network Cameras

Network cameras can be classified in terms of whether they are designed for indoor use only or for both indoor and outdoor use. Outdoor network cameras often have an auto iris lens to regulate the amount of light the image sensor is exposed to. An outdoor camera also will require an external protective housing unless the camera design already incorporates a protective enclosure. Housings are also available for indoor cameras that require protection from harsh environments such as dust and humidity and from vandalism or tampering. (See Chapter 15 for more on enclosures.)

Network cameras — whether for indoor or outdoor use — can be further categorized into the following types:

3.2.1 Fixed Network Cameras

A fixed camera (Figure 3.2) is one whose viewing angle is fixed once it is mounted. A fixed camera with a body and a lens represents the traditional camera type. In some applications, it is advantageous to make the camera very visible. If this is the case, then a fixed camera represents the best choice because both the camera and the direction in which it is pointing are clearly visible. Another advantage is that most fixed cameras have exchangeable lenses. For further protection, fixed cameras can be installed in housings designed for indoor or outdoor installation.

Figure 3.2 Fixed network camera.

Figure 3.3 Fixed dome camera.

3.2.2 Fixed Dome Network Cameras

Fixed dome cameras (Figure 3.3), also called mini domes, essentially consist of a fixed camera preinstalled in a small dome housing. The camera can be easily directed to point in any direction. Its main benefit lies in its discreet, nonobtrusive design, as well as in the fact that it is difficult to see in which direction the camera is pointing. The camera is also tamper resistant. One of the limitations of a fixed dome camera is that it rarely comes with a changeable lens; even if it is changeable, the choice of lenses is limited by the space inside the dome housing. However, a varifocal lens is often provided, which enables adjustment of the camera's field of view. Fixed dome cameras are designed with different types of enclosures such as vandal-resistant or IP66-rated enclosures for outdoor installations. No external housing is required. The mounting of such a camera is usually on a wall or ceiling.

3.2.3 PTZ (Pan, Tilt, Zoom) Network Cameras

PTZ cameras (Figure 3.4) have the obvious benefit of being able to pan, tilt, and zoom through manual or automatic control. In manual operation, an operator can use a PTZ camera to follow, for example, a person in a retail store. PTZ cameras are primarily used indoors and in applications where an operator is employed and where the visibility of the camera's viewing angle is desirable or not an issue. Most PTZ cameras do not have full 360-degree pan and are not made for continuous automatic operation or so-called "guard tours." The optical zoom on PTZ cameras

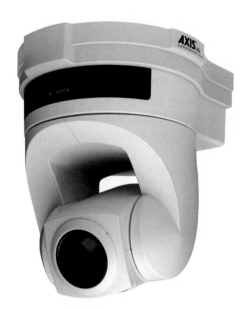

Figure 3.4 Mechanical PTZ camera.

typically ranges from 10X to 26X. A PTZ camera can be mounted on a ceiling or wall.

3.2.4 Nonmechanical-PTZ Network Cameras

A nonmechanical-PTZ camera (Figure 3.5) uses a megapixel sensor and a wide-angle lens to enable it to have a viewing angle of 100 to 180 degrees (or even wider in some cases). Such a camera allows an operator to zoom

Figure 3.5 Nonmechanical-PTZ camera.

25

in on any part of a scene without any mechanical movement. The key advantage is that there is no wear and tear because the camera has no moving parts. Zooming in on a new area of a scene is immediate. In a traditional PTZ camera, this can take up to 1 second. Because a nonmechanical-PTZ camera's viewing angle is not visible, it is ideal for discreet installations. To obtain good image quality, pan, tilt, and zoom should be limited. If such a camera has a 3-megapixel sensor, the recommended maximum viewing angle is 140 degrees with a 3X zoom capability. This type of camera is typically mounted on a wall.

3.2.5 PTZ Dome Network Cameras

PTZ dome network cameras (Figure 3.6) can cover a wide area by enabling greater flexibility in pan, tilt, and zoom functionality. They enable a 360-degree pan and a tilt of usually 180 degrees. Dome cameras are ideal for use in discreet installations due to their design, mounting (particularly in drop-ceiling mounts, as seen in Figure 3.6), and difficulty in seeing the camera's viewing angle (dome coverings can be clear or smoked). A dome network camera also provides mechanical robustness for continuous operation in guard tour mode, whereby the camera continuously moves between presets. In guard tour mode, one PTZ dome network camera can cover an area where ten fixed cameras would be needed. The main drawback is that only one location can be monitored at any given time, leaving the other nine positions unmonitored. The optical zoom typically ranges between 10X and 35X. A PTZ dome network camera often is used in situations where an operator is present. This type of camera is usually mounted on a ceiling if used indoors or on a pole or corner of a building in outdoor installations. With

Figure 3.6 PTZ dome network camera.

a PTZ dome network camera, all PTZ control commands are sent over an IP network, and no RS-485 wires need to be installed, unlike the case with an analog dome camera.

3.3 PTZ Cameras

Although PTZ cameras provide several benefits in video surveillance applications, many times the deployment of such cameras has been cost prohibitive in analog installations. With an analog camera, separate cables are required for video transport (over coax cable) and for sending PTZ commands (over RS-485 cabling) to control the mechanics. With the use of PTZ cameras, substantial savings in installation costs can be achieved because the same network cable is used for transporting both network video and PTZ commands.

As with other kinds of network cameras, today's network dome cameras are surpassing the capabilities of their analog counterparts. One advantage found in some of the latest network dome cameras is the use of a progressive scan sensor that reduces motion blur in an image. This is particularly beneficial in a moving camera, as a moving camera will add blurriness to an image if the traditional interlace scanning technology is used. For more details on interlaced and progressive scanning, see Chapter 4.

3.3.1 Electronic Image Stabilization

For outdoor installations, zoom factors above 20X could be impractical due to vibrations and motion caused by traffic or wind. Vibrations in a video can be reduced if the camera incorporates an electronic image stabilizer. This is useful when the camera is installed in a windy environment, or close to a highway or on a bridge where there are heavy vibrations. Even in an office building, there is normally some level of vibration. Image stabilization is useful even at very high zoom levels.

Image stabilization can be made within the camera's processor — so-called electronic image stabilization (EIS) — as opposed to mechanic image stabilization. This is done through a combination of motion estimation and image shift. This generally requires a slightly oversized sensor and means that the field of view of the output image is smaller than the actual sensor image.

For very rapid movements that occur within the exposure time of the camera, blurring can be reduced by decreasing the exposure time in situations when this is possible. It also is possible to use special lenses that compensate motion during the exposure time.

In addition to the obvious benefit of getting more useful video, using EIS will reduce the file size of the compressed image and lower the bit rate of the video stream, thereby saving valuable storage space (Figure 3.7).

Figure 3.7   Video without EIS (top), and video with EIS (bottom).

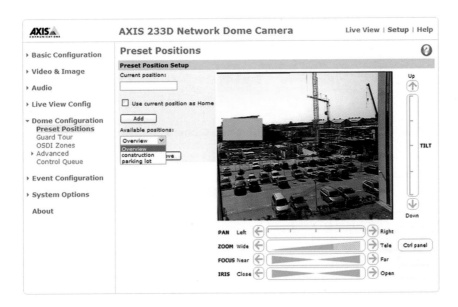

Figure 3.8 Many network dome cameras have as many as 100 preset positions that can be set.

3.3.2 Presets and Guard Tours

Many dome and PTZ cameras have a number of preset positions, normally between 20 and 100 (Figure 3.8). Once the preset positions have been set in the camera, it is very quick for the operator to go from one position to the next. Many network dome and PTZ cameras also have built-in guard tours (Figure 3.9). A guard tour enables the camera to automatically move from one preset position to the next in a predetermined order or at random. The viewing time between one position and the next is configurable. Different guard tours also can be set up and activated during different times of the day.

3.3.3 Privacy Masking

A network dome or PTZ camera can provide surveillance of very large areas, sometimes even in areas where the operator should not monitor. For example, a network dome camera installed outside a football stadium in a city center may likely be able to view and also zoom in on a nearby apartment complex. In such instances, privacy masking becomes important because it allows some areas of a scene to be blocked or masked from viewing and recording (Figure 3.10).

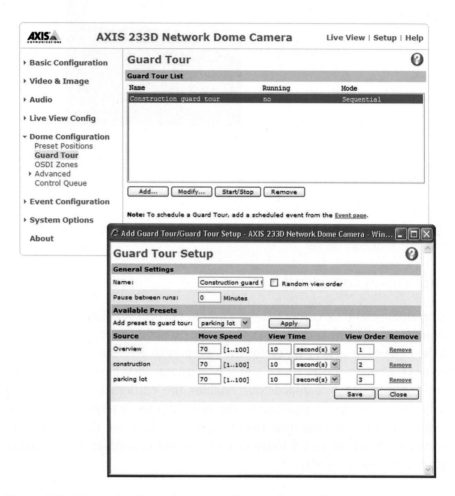

Figure 3.9 Example of a camera interface with guard tour setup. Normally, up to 20 guard tours can be programmed.

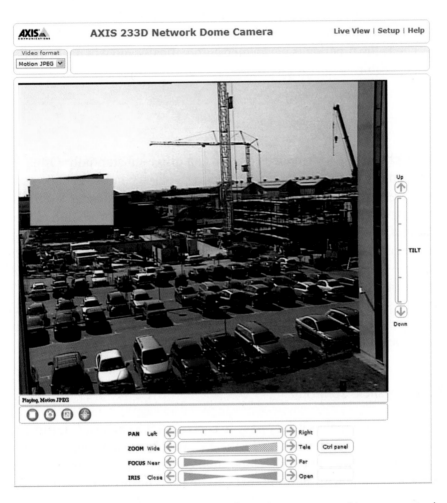

Figure 3.10 With built-in privacy masking (gray rectangle in image), the camera can guarantee privacy for areas that should not be covered by a surveillance application.

3.3.4 E-flip

When a PTZ is mounted on a ceiling and used, for example, to follow a person in a retail store, there will be situations when a person will pass just underneath the camera. An E-flip functionality will electronically rotate the images 180 degrees so that the person and the entire picture will not be upside down when the person passes just underneath the camera. The E-flip function is performed automatically and will not be noticed by an operator.

3.3.5 PTZ Performance

The mechanical performances of PTZ can differ substantially. Often, the maximum performance — given as degrees per second — is indicated in data sheets. If the stated performance of a dome camera is more than 360 degrees per second, it means that the camera can perform a full circle in less than one second. This normally is considered adequate for a high-performance PTZ camera. Some PTZ cameras have even higher performance — upwards of 450 degrees per second. Although high speed is important, so is controllability at very low speeds. Therefore, some PTZ cameras specify the slowest speed with which they can be controlled. For example, a speed of 0.05 degrees per second means it will take two hours to complete just one 360-degree rotation.

3.3.6 Joystick Control

Using a joystick makes it very easy to control a PTZ camera (Figure 3.11). USB joysticks, which connect to the PC used for video monitoring, are commonly used in network video applications. Most professional joysticks also come with buttons that can be used for presets, as described previously.

3.4 Nonmechanical-PTZ Network Cameras

A nonmechanical-PTZ network camera (Figure 3.12) provides the ability to provide both full overview images as well as close-up images using instant zoom, with no moving camera parts. This is achieved by taking advantage of the high resolution provided by a megapixel sensor and a wide-angle lens.

Figure 3.11 A joystick makes controlling PTZ cameras very intuitive and accurate.

Figure 3.12 Nonmechanical-PTZ network camera.

3.4.1 Selecting the Right Viewing Angle

Depending on the type of lens used, viewing angles of up to 180 degrees are achievable. A wide viewing angle, however, comes with some challenges. One challenge is that the image will be greatly distorted by the fish-eye effect from the lens and will need to be de-warped to make it viewable. De-warping an image requires a lot of processing power in the camera or viewing station, which reduces the equipment's performance. The other problem is that only a very limited part of the image sensor can be used, which reduces resolution and thereby image quality. Having, for example, a 140-degree field of view instead of a 180-degree view enables

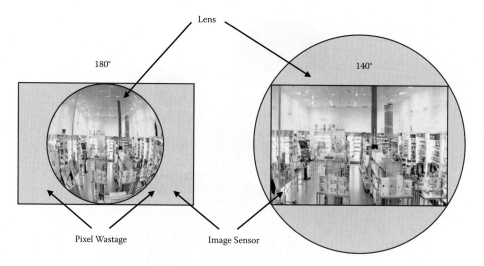

Figure 3.13   With a 180-degree field of view, only part of the sensor is being used. With a 140-degree field of view, the entire sensor can be used for the image.

the use of the full breadth of the image sensor. This means that more of the horizontal area of a scene — the most important area in many applications — is covered, whereas the top and bottom areas of a scene are reduced (see Figure 3.13).

3.4.2 Typical Nonmechanical-PTZ Camera

A nonmechanical-PTZ network camera that uses a 3-megapixel sensor is able to capture an image measuring 2048 × 1536 pixels. Such a camera provides an overview image, which is slightly distorted due to the lens, and zoomed-in images of any part of the overview image. When the camera is instructed to make a 3X zoom in any area of the full overview image, the original megapixel resolution is used to provide a full 1:1 ratio in VGA resolution. The resulting close-up image offers good details with maintained sharpness. This is in contrast to a normal digital zoom, where the zoomed-in image often loses detail and sharpness (see Figure 3.14).

3.4.3 Comparison with a PTZ Camera

A nonmechanical PTZ network camera offers a wider field of view than a mechanical PTZ camera, and there is no wear-and-tear due to moving parts. It is, however, similar to a mechanical PTZ or dome camera in that

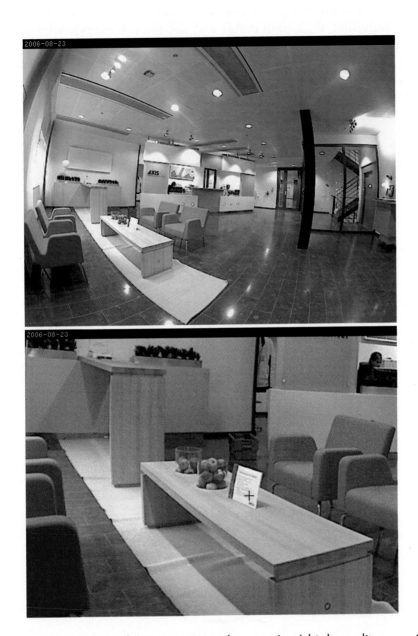

Figure 3.14 High-quality overview and zoom: (top) high-quality overview image in VGA resolution; and (bottom) when making a 3X zoom, a normal VGA 1:1 resolution results.

what one sees is what one records; that is, if an operator zooms in on a certain area of an image, the nonmechanical PTZ network camera can record only that zoomed-in image and not the full overview image (see Figure 3.15).

Figure 3.15 A conventional PTZ camera can cover a bigger area but it can only see a small part at the same time (top), whereas a nonmechanical-PTZ camera sees the entire monitored area (bottom).

3.5 Day and Night Network Cameras

All types of cameras (e.g., fixed cameras, fixed dome cameras, and PTZ cameras) can offer day and night functionality. A day and night camera is designed for use in outdoor installations or in indoor environments with poor or no lighting. The day and night functionality can be achieved in two different ways. The simple way is to decrease the chrominance (color) sensitivity of the camera, thereby turning it into a black-and-white camera — which improves somewhat the camera's light sensitivity. This is, however, not a true day and night functionality; nonetheless, it is marketed by some vendors as such.

A true day and night functionality is only achieved if a mechanically removable IR-cut filter, also called IR-blocking filter, is used in the camera.

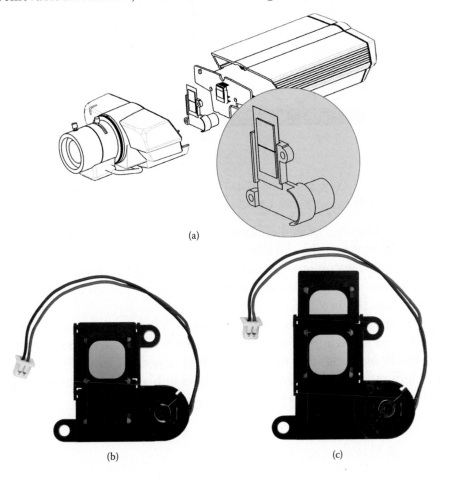

(a)

(b)   (c)

Figure 3.16 (a) IR-blocking filter in a day and night network camera; (b) IR filter, daytime; and (c) IR filter, nighttime.

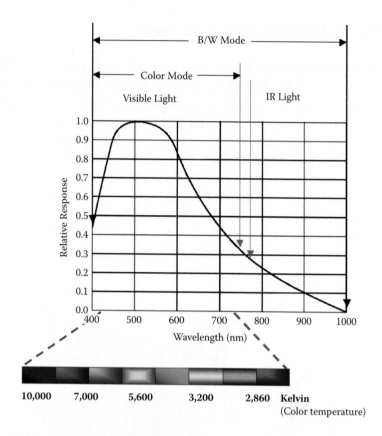

Figure 3.17 By removing the IR-cut filter, the camera's image sensor will become sensitive to near-IR wavelengths.

When the filter is removed, it allows the image sensor to become sensitive to near-IR wavelengths and enables the camera's light sensitivity to reach down to 0.001 lux or lower. In most true day and night cameras, the IR-cut filter is removed automatically when the lighting is below a certain level (e.g., 1 lux) (Figure 3.16).

A day and night color network camera delivers color images during the day. As light diminishes, the camera switches to night mode, removing the IR blocking filter to take advantage of near-IR light (Figure 3.17). In night mode, the camera delivers video in black and white to reduce noise (graininess) and to provide clear, high-quality images.

3.5.1 IR Illuminators

IR illuminators should be used if the goal is to conduct 24/7 surveillance in areas with low light. Many different types of IR illuminators are

Figure 3.18 An IR illuminator.

available (Figure 3.18). Some produce light at a low wavelength, about 850 nm, which will display a faint red hue that is visible to the human eye if the room is completely dark. Totally covert IR illuminators with higher wavelengths of about 950 nm also are available. The downside of higher wavelengths is that they have a substantially shorter reach. The illuminators also come in different illumination angles, for example, 20, 30, or 60 degrees. A wide-angle illuminator normally means a shorter distance will be illuminated. Most IR illuminators today use LEDs (light emitting diodes), which provide a cost-efficient solution that will enable the illuminators to last between 5 and 10 years. Some cameras even come with a built-in IR lamp but the range of the illuminator is normally quite limited. Figure 3.19 provides images without and with an IR illuminator.

3.5.2 Day and Night Applications

Day and night or IR-sensitive cameras (Figure 3.20) are useful in certain environments or situations that restrict the use of artificial light. They include low-light video surveillance applications, where light conditions are less than optimal; covert surveillance situations; and discreet installations, for example, in a residential area where one does not want bright lights to disturb residents at night. An IR illuminator that provides near-infrared light also can be used in conjunction with a day and night or IR-sensitive camera to further enhance the camera's ability to produce high-quality video in low-light or nighttime conditions.

3.6 Megapixel Network Cameras

Megapixel network cameras incorporate a megapixel image sensor that enables the delivery of images with more than one million pixels, usually

Figure 3.19 Image (top) without an IR illuminator and (bottom) with an IR illuminator.

Figure 3.20 An IR-sensitive network camera with an IR illuminator attached below the camera housing.

at a minimum resolution of 1280×1024 pixels. This is four times better pixel resolution than that provided by fixed network cameras with VGA resolution and more than three times better pixel resolution than what can be provided by analog cameras.

3.6.1 The Benefits of Megapixel

A megapixel fixed network camera can be used in one of two ways: (1) it can enable viewers to see greater details (have higher resolution) in an image, or (2) it can be used to cover a larger part of a scene if the image scale (in pixels per area) is kept the same as a non-megapixel camera. See Figures 3.21 through 3.23.

Figure 3.21 Images using a network camera with VGA 640×480 resolution (at top); close-up section (at bottom).

Figure 3.22 Images using a network camera with 1280×1024 or 1.3-megapixel resolution. Same area of coverage. More details can be seen in an image from a megapixel camera because it offers three to four times higher pixel resolution than an image from a non-megapixel camera (Figure 3.21).

Figure 3.23 Image with VGA 640×480 resolution (top); image with 1280×1024 or 1.3-megapixel resolution (bottom). Higher resolution cameras provide wider coverage, with the same pixel resolution.

3.6.2 Megapixel Applications

Megapixel network cameras are used in a number of key industry segments to better address some video surveillance challenges, such as:

- *In retail.* A drastic reduction in theft and shrinkage can result from effective video surveillance. Megapixel network cameras can either provide an overview of a large part of a store (without any blind spots) or offer highly detailed images of the sales counter area.
- *In city surveillance.* Megapixel network cameras provide high-resolution video streams from locations where it is necessary to clearly identify people and objects or get a larger overview while viewing live or recorded video.
- *In government buildings.* Megapixel network cameras provide the exceptional image detail necessary to facilitate the identification of people and to record evidence of any suspicious behavior.
- *In school buildings.* The use of megapixel resolution cameras in hallways makes it easier to identify students.

Megapixel resolution is currently offered in most types of network cameras (Figure 3.24), including fixed, fixed dome, and wireless cameras. The most common resolutions are 1.3 and 2 megapixels. Analog CCTV cameras cannot offer megapixel resolution due to their restrictions to traditional TV line standards. (For more on resolutions, see Chapter 4.)

Figure 3.24 Megapixel cameras today come in many different types, including fixed and fixed dome cameras.

3.6.3 The Drawbacks of Megapixel

Although megapixel resolution enables either higher details or more of a location to be seen, there are drawbacks to consider with current megapixel technology. Megapixel cameras are normally less light sensitive than non-megapixel network cameras. For manufacturing and cost reasons, many megapixel sensors are the same size or only slightly larger than non-megapixel sensors. This means that while there are more pixels on a megapixel sensor, the size of each pixel is smaller than the size of each pixel on a non-megapixel sensor. The smaller the pixel size, the lower the light-gathering ability. As a result, megapixel network cameras are generally less light sensitive than non-megapixel network cameras.

Many megapixel cameras today do not have the ability to provide 30 frames per second at full resolution. Another factor to consider when using megapixel network cameras is that higher-resolution video streams increase demands on network bandwidth and storage space for recorded video. This can be mitigated somewhat using the MPEG-4 video compression standard. However, not all megapixel network cameras support MPEG-4 compression; many support only Motion JPEG. Megapixel cameras that support the H.264 compression standard will be even more useful because the compression enables lower bandwidth and storage use. (For more on compression, see Chapter 5.)

3.7 Best Practices

In rapidly growing markets such as the network camera market, many new vendors are entering the market with new products. There are currently more than 200 different network camera vendors. Because network cameras include much more functionality than analog cameras, choosing the right camera becomes not only more important, but also more difficult.

Considerations to make include:

- *Camera type.* Should it be a fixed, fixed dome, or a pan/tilt/zoom camera?
- *Indoor or outdoor environment.* Where will the camera be located? In an outdoor environment, an auto iris lens and better light sensitivity may be required.
- *Image quality.* How important is image quality? The quality differs between different cameras. Ensure the selection of a camera with the appropriate quality.

- *Resolution.* Will VGA suffice, or is megapixel resolution required?
- *Intelligence.* Is built-in intelligence, such as tamper alarm, a requirement?
- *Network functionality.* Does the chosen network camera have all the appropriate protocol support required in today's demanding enterprise networking?
- *Vendor.* With more than 200 brands on the market, will the camera brand selected be on the market years from now, and does it have the appropriate quality and warranty?

Chapter 4

Camera Technologies

Cameras are complex mechanical and electrical devices based on many different technologies. Understanding these technologies is important to make the most of a video surveillance system.

One of the most important features — if not the most important — of any camera is image quality. This especially is true in video surveillance, where lives and property may be at stake. But how is image quality defined, and how can it be measured and guaranteed?

Image quality can be defined in at least three ways: (1) how aesthetically pleasing an image is to the eye, (2) how well it reflects reality, and (3) how well it meets the aim of a viewing application (e.g., the aim may be to capture a good overview of events taking place or enable identification of a person or object). Although meeting all three criteria well may not be achievable, most would agree that having a clear, sharp, and correctly exposed image that delivers critical information is important in a surveillance application. How well a network camera delivers on this depends on many factors. To understand them, one must understand the process and the elements that influence image generation.

This chapter provides a discussion of light, lens, and image sensors as they relate to network cameras, scanning techniques (progressive and interlaced), image processing (including wide dynamic range), and resolution. Chapter 5 discusses compression, which also relates to image quality.

4.1 Light

Light plays a critical role in determining the quality of an image. This section discusses the characteristics and properties of light and how they affect image quality. It also provides discussion on illuminance and what a camera's lux measurement means. In addition, it explains how cameras can take advantage of near-infrared light to produce good-quality, black-and-white images in low-light environments.

4.1.1 Light Characteristics

Visible light comes in different forms, different directions, and different color hues — all of which affect image quality.

Some common forms of light in a scene are:

- *Direct light* from a point source or small bright object (e.g., sunlight or spotlight) creates sharp contrasts with highlights and shadows. Diffuse light is light from a source that is so much larger than the subject that it illuminates the subject from several directions (e.g., gray sky, an illuminated screen, diffuser or light bouncing off a ceiling).
- *Diffuse light* lowers contrasts, which affects the brightness of colors and the level of detail that can be captured.
- *Specular reflection* is light from one direction bouncing off a smooth surface and being reflected in another direction (e.g., reflection off water, glass, or metal). Specular reflections within an image can be a problem and can reduce visibility. Such reflections can sometimes be reduced using a polarizing filter in front of a camera lens.

It also is important to consider the directions of the light sources in relation to the subject. Light direction is a factor in determining how much detail can be obtained from an image. The following are the main light directions:

- *Frontal light (the light source that comes from behind the camera).* The scene in front of the camera is well illuminated. This is the ideal lighting situation.
- *Sidelight.* This may create great architectural effects but also will produce shadows.

- *Backlight (straight into the camera lens).* This light direction is difficult to handle. Silhouettes of objects can be created, and details and color can be lost.

To manage difficult light situations, try to avoid backlight or add artificial light sources. In most indoor installations where backlight or reflection is present — for example, light from large windows — add frontal lighting, if possible. Use diffusers or reflectors to create good illumination. For a discussion on how cameras handle backlight situations and scenes with complex or high-contrast lighting conditions, see Section 4.5 on image processing.

Figures 4.1a through 4.1g are images that illustrate the effects of different light directions.

(a)

Figure 4.1 (a) Frontal sunlight: details and colors emerge; (b) backlight: details and colors are lost as the camera is placed on the opposite side of the gas station; (c) Problem A: Windows can create reflections and bad exposure; (d) blinds might help: reflections are gone but image is underexposed; (e) solution to Problem A: additional frontal lighting — details appear with added frontal light and correct exposure; (f) Problem B: partial backlight — non-identifiable text on vending machine; (g) solution to Problem B: additional frontal lighting (text on vending machine appears with added frontal light). Continued.

(b)

(c)

Figure 4.1 Continued.

(d)

(e)

Figure 4.1 Continued.

(f)

(g)

Figure 4.1 Continued.

4.1.2 Illuminance

As discussed, different forms and different directions of light affect image quality. In general, the more light on the subject, the better the image. With too little light, focusing will be difficult and the image will be noisy or dark.

How much illuminance is required to produce a good-quality image depends on the camera. A network camera's light sensitivity often is specified in terms of lux, which corresponds to a level of illuminance in which a camera produces an acceptable image. The lower the lux specification, the more light sensitive the camera. Normally, at least 200 lux is needed to illuminate an object so that a good-quality image can be obtained. A high-quality camera can be specified to work down to 1 lux, which means that one can capture an image at 1 lux but it may not be of very high quality.

4.1.2.1 The Definition of Lux

Lux (also known as lumen per square meter or meter-candle) is the amount of light falling onto a surface per square meter. In the lux scale, one lux is equal to the amount of light falling on a one-square-meter surface that is one meter away from a candle. Correspondingly, 10 lux is the amount of light measured at a distance of one meter from ten candles. Foot-candle is another unit for illuminance. One foot-candle is equal to 10.7 lux.

Different light conditions offer different illuminances. Surfaces in direct sunlight receive 100,000 lux, whereas surfaces in full moonlight receive 0.1 lux. Many natural scenes have fairly complex illumination, with both shadows and highlights that give different lux readings in different parts of a scene. Therefore, it is important to keep in mind that one lux reading cannot be an indication of the light condition for a scene as a whole. A lux measurement (using a lux meter, which is a tool that reads the illuminance level) must be specified for a particular surface.

Figures 4.2a through 4.2e show samples of environments with a lux reading for a specific area of a scene.

It is also important to note what a lux meter does not measure. A lux meter does not take into account the amount of light reflected from an object. For example, in a scene with two objects placed side by side — one dark and one light-colored — a lux reading on the surface of the dark object is the same as on the light-colored object although the dark object reflects less light than the light-colored object. Because it is the amount of light reflected from objects that a camera records, it is important to note that the actual amount of light being captured by a camera may be lower or higher than a particular lux reading due to the reflectance factor. A lux meter also only measures visible light. See Figure 4.3.

4.1.2.2 Lux Rating of Network Cameras

Many manufacturers specify the minimum level of illumination needed for a network camera to produce an acceptable image. Although such specifications are helpful in making light-sensitivity comparisons for cameras

(a)

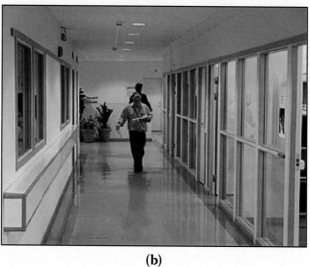

(b)

Figure 4.2 (a) Ground illuminated by 5 lux; (b) office floor illuminated by about 150 lux; (c) shopping mall floor illuminated by about 500 lux; (d) sunny morning and the building is illuminated by about 4,000 lux; and (e) examples of various levels of illuminance. Continued.

(c)

(d)

Figure 4.2 Continued.

Illuminance	Example
0.00005 lux	Starlight
0.0001 lux	Moonless overcast night sky
0.01 lux	Quarter moon
0.1 lux	Full moon on a clear night
10 lux	Candle at a distance of 30 cm (1 ft)
50 lux	Family living room
150 lux	Office
400 lux	Sunrise or sunset
1,000 lux	Shopping mall
4,000 lux	Sunlight morning
32,000 lux	Sunlight at midday (min)
100,000 lux	Sunlight at midday (max)

(e)

Figure 4.2 Continued.

Figure 4.3 A lux meter is a tool that can be used to better understand lighting and its effect on image quality.

produced by the same manufacturer, it may not be helpful to use such numbers to compare cameras from different manufacturers (Figure 4.4). This is because different manufacturers use different methods and have different criteria for what is an acceptable image. An additional complexity is measuring the light sensitivity of day and night cameras. Very low lux values often are specified for such cameras due to their sensitivity to near-IR light. Lux, however, is only defined for visible light, so it is not totally correct to use it for IR-sensitivity, although it is commonly done.

4.1.2.3 Lux Rating of Analog versus Network Cameras

When measuring or comparing the low-light performance of an analog and a network camera, there are a few differences to consider. The primary difference is the fact that the output signal is either analog (a voltage) or digital (a number). There are also differences in how the signal from the sensor is processed as well as some differences in terminology.

Figure 4.4 Images were taken with two similarly priced network cameras from brand-name vendors under a lighting condition of 1 lux. The image on the bottom clearly looks better. The image on the top, however, was from a camera that was specified to have the lower light sensitivity level, 0.8 lux versus 1 lux.

Analog video signals are often described in terms of IREs (from 0 to 100) and digital signals in digital units (often from 0 to 255), with the lowest value meaning black (empty pixel) and the highest, white (full pixel). IRE is a unit defined by the Institute of Radio Engineers. The 100 IRE span corresponds approximately to a voltage difference of 0.7 volts in the video signal, or 1.4 volts peak-to-peak. There are some minor differences between the NTSC and PAL video systems. In NTSC, for example, 7.5 IRE is defined as black, whereas 0 IRE is used for black in PAL. When "0.8 lux at 30 percent IRE" is defined as the low-light performance in a datasheet, it means that at 0.8 lux, only 30 percent of the analog voltage can be detected on the BNC output of the analog camera (i.e., 0.21 volts, or 0.42 volts peak-to-peak).

The signal in both analog and network cameras increases with increasing exposure to light. Light exposure can be achieved by increasing the exposure time or by opening the iris. However, if one uses the maximum exposure time and the largest iris and still wants a brighter image, the signal must be amplified. A higher gain means a brighter image but often also a noisier one because both signal and noise are increased together.

To properly compare the low-light performance of two different cameras, it is necessary to look beyond lux and IRE numbers. Putting two cameras side by side and comparing the outcome is recommended.

4.1.3 Color Temperature

Another consideration to keep in mind is how different types of light affect the color of images. Many types of light (e.g., sunlight and incandescent lamps) can be described in terms of their color temperature (Figure 4.5), which is measured in degrees Kelvin (K). The color temperature scale is based on the fact that all heated objects radiate. The first visible light radiating from a heated object is red. As the temperature of the heated object rises, the radiating color becomes bluer. Red has a lower color temperature than blue. (Notice that this is just the opposite of what is sometimes meant with the colors blue and red when indicating hot or cold water.) Near dawn, sunlight has a low color temperature (implying

Visible Light

Color Temp. 10,000 K   5,500 K   2,860 K

Figure 4.5 Color temperature of visible light from highest to lowest.

redder colors), whereas during the day, it has a higher one (more yellow, neutral colors).

At midday, sunlight has a color temperature of about 5,500 Kelvin, which also is about the temperature of the solar surface. Meanwhile, a tungsten light bulb has a color temperature of about 3,000 Kelvin. Some light sources, such as fluorescent lamps that are gas-discharge lamps rather than heated filaments, are further from the approximation of a radiating heated object. Such light sources cannot be described as accurately in terms of degrees Kelvin. Instead, the closest color temperature is used.

Figure 4.6 shows scenes illuminated by different light sources.

Scenes illuminated by light sources with different colors (or spectral distribution) look different to a camera. The human visual system, however, has a remarkable way of coping with such changes so that colored objects appear to maintain their color. This is sometimes referred to as color constancy. For a camera to do the same thing, it must adapt to the

(a)

Figure 4.6 Scenes illuminated by different light sources. (a) Reception: indoor office lights. Standard light bulbs (3,000 K) are more reddish than daylight and create a brown or yellow tone in the image. To compensate, the camera uses white balance techniques. (b) Warehouse: industrial long-tube fluorescent lights are designed to offer unobtrusive light. However, the images will appear with a green and dull tone. Images need white balance compensation. (c) Shopping mall; some artificial electrical light may offer something similar to daylight. (d) Sunlight at midday; different colors depending on the time of day. Continued.

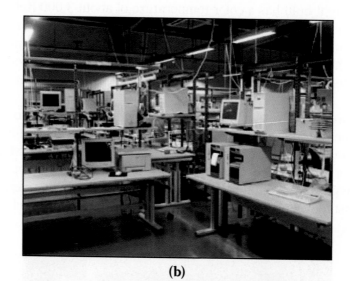

(b)

(c)

Figure 4.6 Continued.

(d)

Figure 4.6 Continued.

local illumination. This process is sometimes referred to as white balancing. In its simplest form, it uses a known object (usually gray) and makes color adjustments to an image so that a gray color (and all other colors) in a scene appears as the human visual system perceives it.

Most modern cameras have an automatic white balance system. White balance often can be adjusted manually or selected among presets.

4.1.4 Invisible Light

As discussed previously, the color (or spectral distribution) changes when the temperature of the light source changes. In the range from just below 3,000 to 10,000 Kelvin, the colors are still in the visible range and, thus, we can see them. For cooler or hotter objects, the bulk of the radiation is generated within the invisible wavelength bands.

Outside the visible range of light, we find infrared (IR) and ultraviolet (UV) light, which cannot be seen or detected by the human eye. Most camera sensors, however, can detect some of the near-infrared light, from 700 nanometers up to about 1,000 nanometers, which can distort the color of resulting images if such light is not filtered out (Figure 4.7). Therefore, a color camera is outfitted with a filter, which is an optical

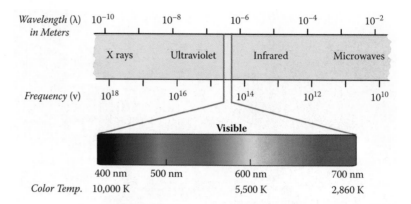

Figure 4.7 Light wavelengths. Near-infrared light spans the 700- to 2,500-nanometer range.

piece of glass placed between the lens and the image sensor. The IR filter will filter out near-IR light and deliver the same color interpretations that the human eye produces. UV light, on the other hand, does not affect a surveillance camera because CCD (charge-coupled device) and CMOS (complementary metal–oxide–semiconductor) sensors are not sensitive to this light. An analog film camera, however, is sensitive to UV (UVA) light and requires a UV filter or coating on the lens of the camera.

The IR blocking filter, commonly called the IR-cut filter, can be removed to extend a network camera's ability to produce quality images in low-light or dark situations. This allows a camera's image sensor to use the near-IR light to deliver high-quality black-and-white images. Cameras with the ability to make use of near-IR light often are marketed as day and night cameras or IR-sensitive cameras. This does not mean that such cameras produce heat-sensitive infrared images. Infrared images require specialized infrared cameras that detect far-infrared light (heat) radiating from animate and inanimate objects, where warmer objects, such as people and animals, stand out from typically cooler backgrounds. Such infrared cameras often are used by the military, and are called thermal cameras.

4.2 Lenses

The first camera component used in capturing images is the lens; therefore, it is an important determinant of image quality. A camera lens normally consists of an assembly of several lenses. The lens performs several functions that affect image quality. They include:

- *Defining the field of view,* that is, defining how much of a scene and the level of detail will be captured.
- *Controlling the amount of light* passing through to the image sensor so that an image is correctly exposed.
- *Focusing* by adjusting either elements within the lens assembly or the distance between the lens assembly and the image sensor.

This section discusses different types of lenses, as well as lens characteristics such as field of view, iris, f-number, focusing, mounts, and lens quality.

4.2.1 Lens Types

There are three main types of lenses:

1. *Fixed lens.* With such a lens, only one field of view (it could be a lens offering normal, telephoto, or wide-angle view) is available because the focal length is fixed (e.g., at 4 mm). See Figure 4.8.
2. *Varifocal lens.* This type of lens offers a range of focal lengths and, hence, different fields of view. The field of view can be manually adjusted. Whenever the field of view is changed, the user must refocus the lens manually. Varifocal lenses for network cameras often provide focal lengths that range from 3.0 to 8 millimeters. See Figure 4.9.

Figure 4.8 Fixed lens.

Figure 4.9 Varifocal lens.

3. *Zoom lens.* Zoom lenses are like varifocal lenses in that they enable the user to select different fields of view. However, with zoom lenses, there is no need to refocus the lens if one changes the field of view. Focus can be maintained within a range of focal lengths, for example, 6 to 48 millimeters. Lens adjustments can be either manual or motorized for remote control. When a lens states, for example, 3X zoom capability, it is referring to the ratio between the lens's longest and shortest focal lengths.

4.2.2 Lens Mount Standards

There are two main lens mount standards used on network cameras: (1) the CS-mount and (2) the C-mount. They both have a 1-inch thread and look the same. What differs is the distance from the lenses to the sensor when fitted on the camera:

1. *CS-mount.* The distance between the sensor and the lens should be 12.5 millimeters.
2. *C-mount.* The distance between the sensor and the lens should be 17.526 millimeters. A 5-millimeter spacer (C/CS adapter ring) can be used to convert a C-mount lens to a CS-mount lens (Figure 4.10).

The initial standard was C-mount, whereas CS-mount is an update to this, allowing for reduced manufacturing cost and sensor size. Today, almost all cameras and lenses sold are equipped with a CS-mount. It is possible to mount an old C-mount lens to a camera body with CS-mount using a C/CS adapter ring. If it is impossible to focus a camera, one probably has the wrong type of lens.

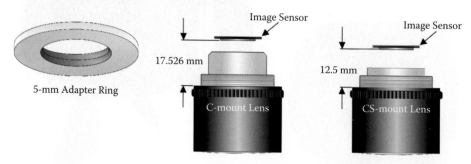

Figure 4.10 C/CS adapter ring (left). C-mount (center), and CS-mount (right).

4.2.3 Field of View (Focal Length)

Images can have normal view (i.e., offering the same field of view as the human eye), telephoto view (a magnification of a narrower field of view, providing, in general, finer details than a human eye can deliver), and wide-angle view (a larger field of view with fewer details than in normal view) (Figure 4.11). It is important to know what field of view one would like the camera to cover because it determines how much, as well as the level of information, one can capture on an image.

The focal length of the lens and the size of its image sensor determine a network camera's field of view. Focal length is defined as the distance between the entrance lens (or a specific point in a complicated lens assembly) and the point where all the light rays converge to a point (normally, the camera's image sensor). The longer the focal length, the narrower the field of view. To achieve a wide field of view, the focal length should be shortened.

To understand terminologies such as field of view; focal lengths; and normal, telephoto, and wide-angle lenses, a traditional camera that uses 35-mm film provides a good base for comparison. The human eye has a fixed focal length that is equivalent to a lens with a focal length of 50

(a)

Figure 4.11 Different fields of view: wide-angle view (a), normal view (b), and telephoto – narrow field of view (c). Continued.

(b)

(c)

Figure 4.11 Continued.

(a)

(b)

(c)

Figure 4.12 (a) Normal lens for a network camera (standard focal length); (b) telephoto lens for a network camera (long focal length); and wide-angle lens for a network camera (short focal length).

millimeters on a traditional 35-mm camera. Hence, traditional lenses that have focal lengths ranging from 35 to 70 millimeters are considered "normal" lenses that provide a "normal" field of view (Figure 4.12a).

Telephoto lenses (Figure 4.12b) for traditional cameras have focal lengths of more than 70 millimeters. A telephoto lens is used when the surveillance object is either small or located far away from the camera. A telephoto lens provides good detail for long-distance viewing and there is normally low geometrical distortion, which appears as curvatures at the edges of an image. However, a telephoto lens has, generally, less light-

Table 4.1 Field of View for a Given Network Camera

Sensor =	Factor ×	Network Camera Lens =	Traditional Camera's Focal Length
(inch)	(inch)	(mm)	(mm)
1/4	9	3–8	27–72
1/3	7	3–8	21–56
1/2	5	3–8	15–40

gathering capability, which requires a scene to have good lighting to produce a good-quality image.

Wide-angle lenses (Figure 4.12c) have focal lengths of fewer than 35 millimeters. The advantages include a wide field of view, good depth of field, and fair low-light performance. The downside is that this type of lens produces geometrical distortions. Lenses with a focal length of fewer than 20 millimeters create what is often called a "fish-eye" effect. Wide-angle lenses are not often used for long-distance viewing.

The focal lengths of a network camera lens are relatively shorter by comparison because the size of an image sensor (the digital equivalent to a 35-mm film) also is much smaller than the size of a frame on a 35-mm film. It is, therefore, not relevant to know about the focal length of a network camera lens unless one also knows the size of a camera's image sensor, which may be of various sizes. Thus, a network camera's field of view is determined by both the lens's focal length and the size of its image sensor.

The typical sizes of image sensors are 1/4, 1/3, 1/2, and 2/3 inch. There are two ways to determine what the field of view is for a given network camera lens. A simple way is to see it in terms of a traditional 35-mm camera. This can be done by making a conversion using the size of the image sensor and its corresponding conversion factor, together with the focal length ranges of a network camera lens. See Table 4.1. The second way is to calculate it using the formula in Figure 4.13. This formula helps determine the height and width of the scene that will be captured, given the focal length of the lens and the distance to the scene or subject from the camera position.

- Calculation – feet:
  What width of objects will be visible at 10 feet when using a camera with a 1/4-inch CCD sensor and a 4-millimeter lens?

$$W = D \times w/f = 10 \times 3.6/4 = 9 \text{ feet}$$

- Calculation – meters:
  What width of objects will be visible at 3 meters when using a camera with a 1/4-inch CCD sensor and a 4-millimeter lens?

$$W = D \times w/f = 3 \times 3.6/4 = 2.7 \text{ meters}$$

When in the field, the fastest way to calculate the lens required and the field of view is to use a rotating lens calculator, which is available from most camera and lens manufacturers (Figure 4.14).

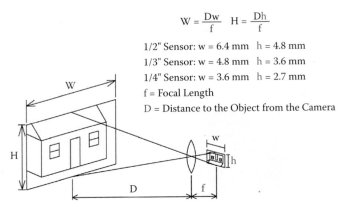

$$W = \frac{Dw}{f} \quad H = \frac{Dh}{f}$$

1/2" Sensor: w = 6.4 mm   h = 4.8 mm
1/3" Sensor: w = 4.8 mm   h = 3.6 mm
1/4" Sensor: w = 3.6 mm   h = 2.7 mm
f = Focal Length
D = Distance to the Object from the Camera

Figure 4.13   In the formula, "w" and "h" correspond to the physical width and height of the image sensor.

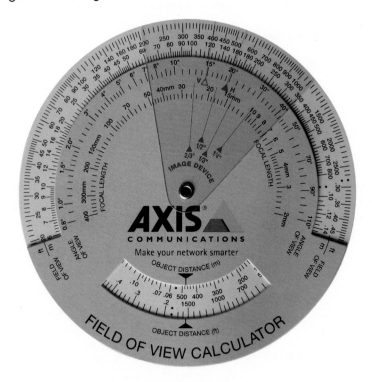

Figure 4.14   A rotating lens calculator is a good tool for quickly calculating the lens required or the field of view.

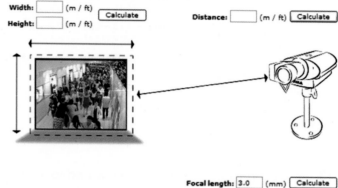

Lens calculator: AXIS 211, AXIS 211A

The focal length of the delivered varifocal lens is 3.0-8.0 mm. Keep this value or enter the focal length of the new lens to find out at what distance you should place the camera in order to capture a specific scene. You can also calculate the focal length of the lens you need by specifying the actual distance and scene dimensions.

Please enter the length values (distance, width or height) specified in the same measurement unit, e.g. meters or feet. The focal length of the lens is always specified in millimeters. Click the button to the right of the parameter you want to calculate.

Width: ☐ (m / ft)   **Distance:** ☐ (m / ft) [Calculate]
[Calculate]
Height: ☐ (m / ft)

Focal length: 3.0 (mm) [Calculate]
Example: If the focal length is 3.0 mm and the distance to the scene is 8 meters, the calculated width and height of the scene will be 9.60 and 7.20 meters, respectively.

Figure 4.15 Online lens calculator.

Several manufacturers also make their lens calculators available online, making calculations quick and convenient (Figure 4.15).

4.2.4 Matching Lens and Sensor

Image sensors are available in different sizes, such as 2/3, 1/2, 1/3, and 1/4 inch, and lenses are manufactured to match these sizes. It is important to select a lens suitable for the camera (Figure 4.16). A lens made for a 1/2-inch sensor will work with 1/2-, 1/3-, and 1/4-inch sensors, but not with a 2/3-inch sensor.

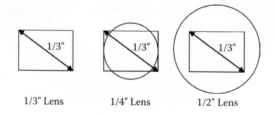

1/3" Lens   1/4" Lens   1/2" Lens

Figure 4.16 Examples of different lenses mounted onto a 1/3-inch sensor.

If a lens is made for a smaller sensor than the one actually fitted inside the camera, the image will have black corners. If a lens is made for a larger sensor than the one actually fitted inside the camera, the field of view will be smaller than the lens's capability because part of the information will be "lost" outside the sensor. It provides a telephoto effect as it makes everything look zoomed in and it is called a focal length multiplier greater than 1. The smaller the image sensor for a given lens, the more a scene is magnified (if resolution is kept the same); see Figure 4.17.

4.2.5 Aperture or Iris Diameter

A lens's ability to pass light through is measured by its aperture or iris diameter. The bigger the aperture of a lens, the more light can pass through it. In environments with low-light conditions, it is generally better to have a lens with a large iris diameter.

To achieve a good-quality image, the amount of light passing through a lens must be optimized. If too little light passes through, the resulting image will be dark. If too much light pases through, the resulting image will be overexposed.

The amount of light captured by an image sensor is controlled by two elements that work together: (1) the size of the lens's iris opening, whereas affects the area light must pass through, and (2) the exposure time, that is, the length of time an image sensor is exposed to light.

The longer the exposure time, the more light an image sensor receives. Bright environments require shorter exposure time, whereas low-light conditions require longer exposure times. A certain exposure level can be achieved using either a large iris opening and a short exposure time or a small iris opening and a long exposure time. It is important to be aware that increasing the exposure time also increases motion blur, whereas increasing the iris opening has the downside of reducing the depth of field, which is explained below. Another parameter that affects the exposure is signal amplification, that is, the adjustment of the gain or sensitivity of the image sensor, which also will increase the noise level.

In a traditional camera, there is a shutter to control the exposure time. However, with most network cameras, no mechanical shutter is used. Instead, the image sensor acts as a virtual electronic shutter, either switching on and off to collect and discharge electrical charges or exposing one line of the sensor at a time (in sequence from top to bottom) in a method called a rolling shutter.

(a)

(b)

Figure 4.17 Field of view for same lens and resolution, but different sensor size: (a) Image (ignore red border) with a 1/3-inch sensor; and (b) image with a 1/4-inch sensor. Because the latter sensor is smaller than 1/3 inch, a smaller part of the scene, as shown bordered in red in (a), is shown. The result is a magnification of a smaller area.

In indoor environments where light levels may be constant, a manual iris lens can be used. This type of lens either provides a ring to adjust the iris, or the iris is fixed at a certain f-number. A lens with an automatically adjustable iris is recommended for outdoor applications and where the scene illumination is constantly changing. The iris aperture is controlled by the camera and is constantly changed to maintain the optimum light level to the image sensor. Most automatic iris lenses are controlled via a standard DC current, controlled by the camera's processor, and are therefore called "DC-iris" lenses.

4.2.6 f-Number (f-Stop)

A lens's focal length and aperture play a key role in determining the quality of an image because they affect the amount of light admitted to the sensor. The bigger the aperture in relation to the lens's focal length, the brighter the image. The relationship is defined as the f-number or f-stop, which is the ratio of the lens's focal length to the diameter of the aperture:

$$f\text{-number} = \text{Focal length}/\text{Aperture}$$

The smaller the f-number (either short focal length relative to the aperture, or large aperture relative to the focal length), the better the lens's light gathering ability; that is, more light can pass through the lens to the sensor.

Often, f-numbers are written as F/x. The slash indicates division. An F/4 means the iris diameter is equal to the focal length divided by 4, so if a camera has an 8-millimeter lens, light must pass through an iris opening that is 2 millimeters in diameter.

In low-light situations, a smaller f-number — that is, better light-gathering ability — generally produces better image quality. (There may be some sensors, however, that are not able to take advantage of a lower f-number in low-light situations due to the way they are designed.) Table 4.2 shows the amount of light relative to the F/5.6 iris position. This means that an F/1.0 lens enables 32 times more light to pass through to the sensor than an F/5.6 lens and, hence, is 32 times more light sensitive.

Table 4.2 Amount of Light Relative to the F/5.6 Iris Position

f-Number	F/1.0	F/1.4	F/2.0	F/2.8	F/4.0	F/5.6
Relative Light Level	32	16	8	4	2	1

Although network cameras with DC-iris lenses have an f-number range, often only the maximum light-gathering end of the range is specified. In addition, a lens with a lower f-number is normally more expensive than a lens with a higher f-number.

4.2.7 Depth of Field

A criterion that may be important to a video surveillance application is depth of field. Depth of field refers to the distance in front of and beyond the point of focus where objects appear to be sharp simultaneously. A camera, for example, may be monitoring a parking lot and one may want the ability to analyze the license plates of cars at 20, 30, and 50 meters (60, 90, and 150 feet) away. See also Figure 4.18.

Depth of field is affected by three factors: (1) focal length, (2) iris diameter, and (3) distance of the camera to the object. The longer the focal length or the shorter the distance of the camera to the object, the shorter the depth of field. Depth of field is also shorter the bigger the iris diameter is. Thus, the smaller the iris opening, the better.

Figure 4.19 provides an example of the depth of field for different f-numbers with a focal distance of 2 meters (7 feet). A large f-number (smaller iris opening) enables objects to be in focus over a longer range.

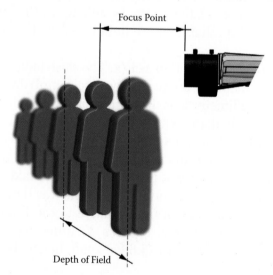

Figure 4.18 Depth of field. Imagine a line of people standing behind each other. If the focus is in the middle of the line and it is possible to identify the faces of all in front and behind the midpoint more than 15 meters (45 feet) away, the depth of field is good.

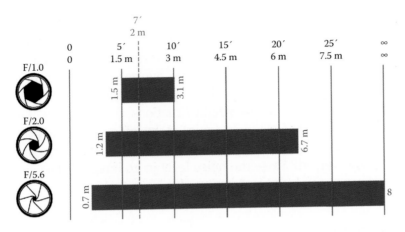

Figure 4.19 Iris opening versus depth of field.

Table 4.3 What Is Achievable under Various Lighting Conditions

	Sunshine	Low-Light Setting 1	Low-Light Setting 2
Exposure Time	Short	Long	Short
Iris Opening	Small (big f-number)	Small (big f-number)	Large (small f-number)
Gain or Sensitivity	Limited	Limited	Full
Image Priority	All	Depth of field (DOF)	Movement
Result	Good DOF, clear image, frozen action[a]	Good DOF, clear image, blurry movement[a]	Limited DOF, noise, frozen action

[a] A very small iris may decrease sharpness.

In low-light situations, there is a trade-off between achieving good depth of field and reducing blur from moving subjects. To get a good depth of field and a properly exposed image in low light, a small iris and long exposure time are necessary. However, the long exposure time increases motion blur. A smaller iris generally will increase the sharpness because optical errors will be reduced, but too small an iris will blur the image due to another optical effect. Getting the optimal result depends on the application requirements and priorities. (See Table 4.3.)

The following are some guidelines of what is achievable under various lighting conditions:

- *Sunshine.* With lots of light, the camera will choose a small iris opening and a short exposure time, providing maximum image quality with good depth of field and clear image without any motion blur.

- *Overcast.* If depth of field is a priority, one needs to increase the exposure time to reduce the iris opening, but note that blurring from a moving subject will increase. For clearer images of moving subjects, reduce the exposure time, but note that depth of field will be compromised with a bigger iris opening.
- *Evening and night.* In lower-light situations, the camera will need to adjust the light sensitivity (gain) of the image sensor to provide good images. Increasing the gain or sensitivity of the image sensor, however, will raise the noise level in an image. Noise can be seen, for example, as grainy particles in open skies.

4.2.8 Focusing

Focusing a network camera often requires making very fine adjustments to the lens. With some auto iris network cameras, one might have problems focusing if one tries to set it under very bright conditions or in dark conditions (not enough light). It often is recommended that the focus of an auto iris lens is adjusted in low-light conditions or with the use of a special dark focus filter. In low-light conditions, the iris automatically opens, resulting in a shorter depth of field, and this helps the user focus more precisely. Setting the focus in bright conditions may be easier, given a longer depth of field, but when light decreases and the iris diameter increases, the image may no longer be in focus due to a shorter depth of field. Due to many different lens designs, it is advisable to follow a camera's installation guide when making lens adjustments for optimal performance.

Some cameras come equipped with an auto focus feature. This means that the camera will automatically adjust the lens mechanically so that a focused image is achieved. The auto focus feature is a requirement in a pan/tilt camera where the camera direction is constantly changing. Some fixed cameras also have the auto focus feature, although it is normally difficult to justify the additional cost because the focus for a fixed camera normally is set only once when it is installed.

4.2.9 Lens Quality

The nature of light limits a lens's so-called resolving power (how small details can be imaged) due to a phenomenon called diffraction. The bigger the lens, or the larger the aperture, the better the resolving power. All lens designs are compromises that are optimized for a particular task.

The quality of the lens material, the coatings on a lens, and how the lens assembly is designed influence the kind of resolution (or how fine a detail) a lens can provide.

Some lenses are made of glass and some of plastic. Although glass lenses are generally better, there is no guarantee that this is always the case. What is more important is a lens's properties and the way it is designed. No lens is perfect, and all create some form of aberration or image defects as a result of the limitations. Some of the aberrations include:

- *Spherical aberration.* Light passing through the edges of a lens is focused at a different distance than light striking near the center of a lens.
- *Astigmatism.* Off-axis points are blurred in the radial or tangential direction. Focusing can reduce one at the expense of the other, but it cannot bring both into focus at the same time.
- *Distortion (pincushion and barrel).* The image of a square object has sides that curve in or out.
- *Chromatic aberration.* The position of sharp focus varies with the wavelength.
- *Lateral color.* The magnification varies with wavelength.

4.2.10 Megapixel Lenses

Megapixel resolution puts higher demands on a lens, which means that replacing a lens on a megapixel camera needs some careful consideration. The main reason is that the pixels on the megapixel sensors are much smaller than the pixels on a VGA sensor and, subsequently, demand a higher-quality lens.

It is not only the quality of the transparent material that matters but also the construction of the aperture (i.e., the iris). When light passes through the aperture, it spreads out. This effect is called diffraction and cannot be avoided. It can be limited somewhat with careful construction of the aperture.

It is also important to use the correct lens resolution. A 1/3-inch lens made for a VGA camera often will not be suitable for a megapixel camera. This is because such a lens can resolve details with reasonable contrast only up to a certain number of lines per millimeter. It is best to match the lens resolution to the camera resolution to fully use the camera's capability and to avoid the effect of aliasing. More details about aliasing appear later in this chapter.

4.3 Image Sensors

As light passes through a lens, it focuses on the camera's image sensor. An image sensor consists of many photosites, and each photosite corresponds to a picture element — more commonly known as "pixel" — on an image. Each photosite or pixel area on an image sensor registers the amount of light it is exposed to and converts it into a corresponding number of electrons. The brighter the light, the more electrons are generated.

Image sensors, by themselves, register only the amount of light, interpreted as shades of gray from white to black. Color can be registered using different methods. One method takes advantage of the three primary colors, red, green, and blue, much like the human eye does. Red, green, and blue, when mixed in different combinations, produce most of the colors that humans see. The most common method used is to apply red, green, and blue color filters in a pattern over the pixels of an image sensor, enabling each pixel to register the brightness of one of the three colors of light. (For example, a pixel with a red filter over it will record the red light while blocking all other colors.)

COLOR AND HUMAN VISION

The human eye has three different color receptors: red, green, and blue. A camera is built to imitate, as close as possible, the human visual system. This means that the camera, as with human vision, translates light into colors that may be based not just on one light wavelength but a combination of wavelengths.

When the wavelength of light changes, the color changes. Blue light has a shorter wavelength than red light. However, a combination of wavelengths or a single wavelength can be interpreted by the human visual system as the same color because the system is not able to tell the difference. Hence, yellow light, for example, could be either a single wavelength or a combination of several (for example, green and red). Thus, there are certain different combinations of wavelengths that look the same.

The most popular red, green, and blue color filter arrangement (RGB) is the Bayer array (Figure 4.20), which consists of alternating rows of red-green and green-blue filters. The Bayer array contains twice as many green as red or blue filters because the human eye is more sensitive to green light than to red and blue light. In addition, having more green pixels produces an image that enables the display of finer details than can be accomplished if each of the three colors were equally applied.

Another color registration method is to use the complementary colors — cyan, magenta, and yellow (CMY). Complementary color filters in sensors often are combined with green filters to form a CMYG color array (Figure 4.21).

Figure 4.20 Bayer array color filter laid over an image sensor.

Figure 4.21 CMYG (cyan, magenta, yellow, green) color filter array.

In comparison to the RGB system, the CMY system appears to offer better light sensitivity because the color filters are different. It is, however, often poorer at presenting colors accurately because a higher dynamic range and more processing are needed.

The CMYG color array often is used in interlaced, CCD (charge-coupled device) image sensors, whereas the RGB system primarily is used in progressive scan image sensors. (More information about interlaced and progressive scan is provided in Section 4.4.)

Once an image is composed following exposure, the electrical charges from the pixels are converted into voltage, and with the use of an analog-to-digital converter, they are transformed into a set of numbers. Once an image is formed, it is sent for processing in a stage that determines, among other things, the colors of each individual pixel that make up an image. (Image processing is discussed in Section 4.5.)

When building a camera, there are two main technologies that can be used for the camera's image sensor (Figure 4.22):

Figure 4.22 (Left) CCD sensor and (right) CMOS sensor.

1. CCD (charge-coupled device)
2. CMOS (complementary metal–oxide–semiconductor)

Although they often are seen as rivals, CCD and CMOS sensors have unique strengths and weaknesses that make them appropriate for different applications. CCD sensors are produced using a technology developed specifically for the camera industry, whereas CMOS sensors are based on standard technology already extensively used in memory chips inside PCs, for example. Modern CMOS sensors use a more specialized technology and the quality of the sensors is rapidly increasing.

4.3.1 CCD Technology

In a CCD sensor, every pixel's charge is transferred through a very limited number of output nodes (often one), where the charges are converted to voltage levels, buffered, and sent off chip as an analog signal. The signal is then amplified and converted into a set of numbers using an analog-to-digital (A/D) converter outside the sensor. All the pixels in the sensor can be devoted to capturing light, and the uniformity of the output (a key factor in image quality) is high. As discussed, the CCD has no built-in amplifiers or A/D converters. Many tasks are performed outside the CCD. (See Figure 4.23.)

CCD sensors have been used in cameras for more than 30 years and present many advantageous qualities. Generally, they still offer better light sensitivity and produce somewhat less noise than CMOS sensors. Higher light sensitivity translates into better images in low-light

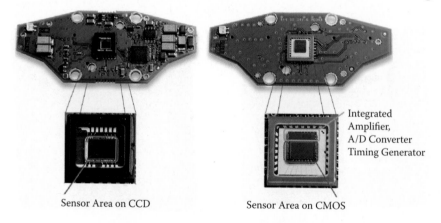

Integrated Amplifier, A/D Converter Timing Generator

Sensor Area on CCD   Sensor Area on CMOS

Figure 4.23 CCD and CMOS sensors mounted on PCBs (Printed Circuit Board). Note the extra components on the CCD-equipped PCB.

conditions. CCD sensors, however, are more expensive and more complex to incorporate into a camera. A CCD also can consume as much as 100 times more power than an equivalent CMOS sensor.

4.3.2 CMOS Technology

The CMOS chip incorporates amplifiers and A/D converter. Individual pixels can be read, allowing "windowing." Windowing allows readout of only parts of the sensor area and can enable high-resolution, digital PTZ, or higher frame rate to be delivered from a limited part of the sensor.

Recent advances in CMOS sensors bring them closer to their CCD counterparts in terms of image quality. CMOS sensors lower the total cost for cameras because they contain all the logics needed to build cameras around them. In comparison with CCDs, CMOS sensors enable more integration possibilities and more functions. They also have lower power dissipation (at the chip level) and a smaller system size. Megapixel CMOS sensors are more widely available and less expensive than megapixel CCD sensors.

4.3.3 More about Image Sensors

In addition to size, resolution, and type of sensors, there are also several other characteristics that differentiate sensors, such as:

- Pixel size
- Light sensitivity of a pixel (including fill factor)
- Maximum signal-to-noise ratio
- Dynamic range
- Fixed-pattern noise

Both the size of a camera's image sensor and the size of each pixel (photosite/silicon square) in the image sensor affect image quality. Image sensors used in most network cameras are usually 1/4 inch or 1/3 inch, which are no more than 4.8 × 3.6 millimeters in size. A larger-sized image sensor has the ability to contain many more pixels than a smaller-sized sensor and can therefore provide higher-resolution images and greater detail.

Also, the larger the size of each pixel in an image sensor, the more capacity each pixel has to store electrons generated from exposure to light. A larger pixel generally means a larger maximum signal-to-noise ratio, with images that are less noisy in highlights.

In addition, a larger pixel often means a higher fill factor and therefore a higher light sensitivity level. The fill factor is the ratio of the area devoted to light gathering compared with the total area, which includes the area devoted to circuitry within a pixel. A sensor with pixels completely devoted to light gathering has 100 percent fill factor. Each pixel on a CMOS sensor has circuitry, and therefore the sensor's fill factor is less.

Thus, when two similar-sized image sensors differ in pixel count and pixel size, they most likely will produce different resolutions and have different light sensitivities. This is an important consideration to take into account when looking at image sensors used in megapixel cameras. See Section 4.3.4 for more information on megapixel sensors.

An image sensor also affects the dynamic range, that is, the range from the maximum useful level to the lowest noise level. To be able to capture both dark and bright objects in the same scene without showing too much noise, the camera's image sensor must have a high dynamic range. Many natural scenes have a rather high range of brightness levels, and sometimes it is difficult for a camera to handle them. Typical examples are indoor pictures with a bright window or an outdoor scene with a dark foreground and a bright sky. There are various techniques that enable a camera to go beyond the limited dynamic range of a typical sensor. They usually work in such a way that individual pixels are exposed or treated differently to reduce noise. Without such techniques, a higher gain on dark pixels also would amplify noise and give a lower-quality image.

The quality of an image sensor also is determined by how much fixed-pattern noise exists. Fixed-pattern noise is noise that has a fixed pattern that does not change over time. It is caused by nonuniformity of the pixels on an image sensor and by electrons from heat generated by the sensor. It is mostly noticeable during long exposures.

4.3.4 Megapixel Sensors

An important consideration to take into account when looking at image sensors used in megapixel cameras is the size of both the image sensor and the pixel. For cost reasons, many megapixel sensors (i.e., sensors containing a million or more pixels) are the same size as or only slightly larger than sensors used in cameras that deliver VGA (Video Graphics Array) resolution of 640×480 (307,200) pixels. The difference is that a megapixel sensor has many more pixels but the pixel sizes normally are smaller than those on a VGA sensor.

For example, a megapixel sensor such as a 1/3-inch, 2-megapixel sensor has pixel sizes measuring 3 micrometers each. By comparison, the pixel size on a 1/3-inch VGA sensor is 7.5 micrometers. This means that although the megapixel camera provides higher resolution and greater detail, it is less light sensitive than its VGA counterpart due to the fact that the pixel size is smaller and light reflected from the same object is spread to more pixels. To maintain a pixel size of 7.5 micrometers in a megapixel sensor would require a sensor that is expensive to produce. As technology advances, it is, however, likely that the performance of sensors utilizing small pixels will improve.

Thus, until such time as we get more light-sensitive and low-noise megapixel sensors at reasonable prices, it will not be uncommon for the arriving generation of megapixel cameras to be less light sensitive and have more noise (although noise reduction algorithms can be used to limit it) than VGA-resolution cameras.

A megapixel sensor can be used innovatively in a nonmechanical-PTZ (pan, tilt, zoom) network camera to provide high-quality, full-overview images or close-up images using instant zoom with no moving camera parts. For more details on this subject, see Section 3.4 on nonmechanical-PTZ network cameras.

4.4 Image Scanning Techniques

Interlaced scanning and progressive scanning are the two techniques available today for reading and displaying information produced by image sensors. Interlaced scanning is used mainly in CCDs. Progressive scanning is used in either CCD or CMOS sensors. Network cameras can make use of either scanning technique. Most analog cameras, however, are based on standards that can only make use of the interlaced scanning technique for transferring images over coax and for displaying them.

4.4.1 Interlaced Scanning

When an interlaced image from a CCD is produced, two fields of lines are generated: (1) a field displaying the odd lines and (2) a second field displaying the even lines. However, to create the odd field, information from both the odd and even lines on a CCD sensor is combined. The same goes for the even field, where information from both the even and odd lines is combined to form an image on every other line.

Interlaced video is an invention from the 1930s. At that time, it was not possible to scan and broadcast reasonably high-resolution video at high speeds because it required too much bandwidth and complexity. The solution was interlacing. Although some network cameras do incorporate interlaced CCD sensors, the interlaced scanning technique primarily is used in analog cameras.

The interlacing technique involves transmitting only half the number of lines (alternating between odd and even lines) of an image at a time. The technique also requires that the monitor be interlaced — presenting every other line and refreshing these fields at a rate that the human visual system interprets as complete images rather than alternating fields of images. This is how a traditional television works. A television set shows video by first displaying the odd lines and then the even lines of an image and refreshes them alternately at 30 (NTSC) or 25 (PAL) frames per second. All analog video formats and some modern HDTV (high-definition television) formats are interlaced. Although the interlacing technique creates artifacts or distortions as a result of "missing" data, they are not very noticeable on an interlaced monitor.

However, when interlaced video is shown on progressive scan monitors such as computer monitors — which scan lines of an image consecutively — the artifacts become noticeable. The artifacts, which can be seen as "tearing," are caused by the slight delay between odd and even line refreshes as only half the lines keep up with a moving image while the other half waits to be refreshed. It is especially noticeable when the video is stopped and a freeze frame of the video is analyzed. For moving objects, interlaced video can provide a more fluid motion in live viewing mode because the image fields are updated twice as often as with progressive scan.

4.4.2 Deinterlacing Techniques

To show interlaced video on computer screens and reduce unwanted interlacing effects (tearing), different deinterlacing techniques can be employed. The problem with deinterlacing is that two image fragments, captured at different times, must be combined into an image suitable for simultaneous viewing. There are several ways to limit the effects of interlacing.

One of the deinterlacing techniques involves removing every other field (odd or even) and doubling the lines of each remaining field (consisting of only even or odd lines) by simple line doubling or, even better, using interpolation. This results in the videos having effectively half the vertical resolution, scaled to the full size. Although this prevents accidental blending of pixels from different fields (the comb effect), it causes noticeable reduction in picture quality and less smooth video. (See Figures 4.24b, d.)

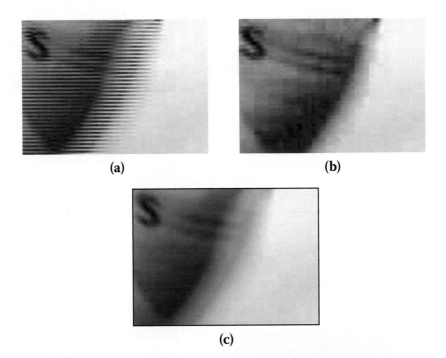

Figure 4.24 (a) Original interlaced image shown on a progressive scan monitor; (b) image deinterlaced with line doubling; (c) image deinterlaced with blending; (d) line doubling technique; and (e) blending technique.

Continued.

Another deinterfacing technique involves blending consecutive fields and displaying two fields as one image. The advantage of this technique is that all fields are present, but the "comb effect" may be visible because both fields, which are captured at slightly different time frames, are simply merged together. The blending operation may be done differently, giving more or less loss in the vertical resolution of moving objects. This often is combined with a vertical resize so that the output has no numerical loss in vertical resolution. The problem with this technique is that there is a loss in quality because the image has been downsized and then upsized. The loss in detail makes the image look softer, and the blending creates the ghosting artifacts. (See Figure 4.24c, e.)

Yet another technique is BOB deinterlacing (because the fields are "bobbed" up and down), whereby each field is made into a full frame and each odd frame is pushed down by half a line and each even frame is pushed up by half a line. This method requires one to know which field, whether odd or even, should be displayed first. The total number of frames for the video is doubled, and to play the video at the right speed, the frame

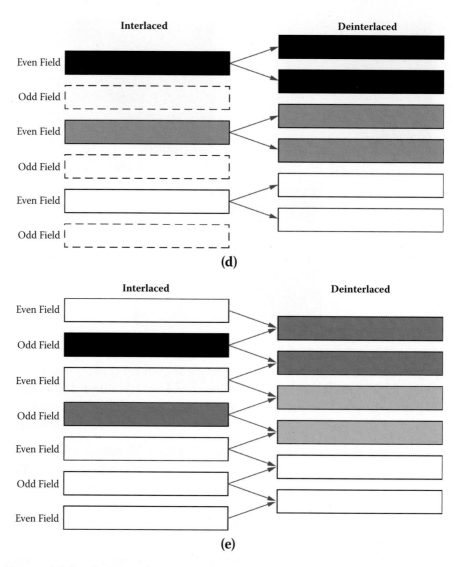

Figure 4.24 Continued.

rate is doubled. This technique leads to some line flickering and requires more computer power to play and store.

Another trivial method of removing interlacing artifacts is to scale down the video horizontally and throw every other field out, thus keeping only half of the information. The video size is smaller and only half of the temporal resolution is retained but no artifacts are visible.

A more advanced technique is motion adaptive deinterlacing. This technique incorporates the previously described technique of blending or averaging, together with a calculation for motion. If there is no motion, the two consecutive fields are combined to form a complete frame, and full resolution, as well as sharpness, is obtained. When motion is detected, vertical resolution is compromised as before. In pixel-based motion adaptive deinterlacing, the nonmoving parts of an image are shown in full resolution while pixels that cause artifacts in the moving parts of the image are discarded. Lost data at the edges of moving objects then can be reconstructed using multi-direction diagonal interpolation. This technique requires a substantial amount of processing power.

Interlaced scanning has served the analog camera, television, and VHS video world very well for many years and is still the most suitable technique for certain applications. However, with progressive computer monitors, the advent of liquid crystal display (LCD), thin-film transistor (TFT)-based monitors, DVDs, and digital cameras, an alternative method of bringing images to the screen known as progressive scanning is becoming more widely used. See Figure 4.25.

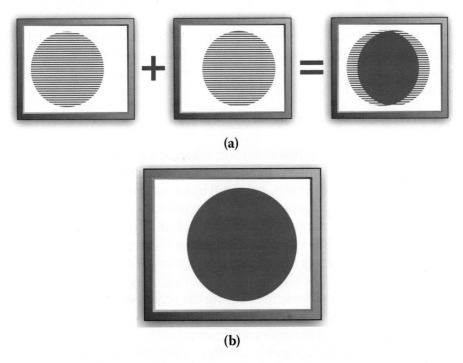

(a)

(b)

Figure 4.25 (a) An interlaced scan image shown on a progressive (computer) monitor. (First field: odd lines; second field: even lines, 17/20 milliseconds (NTSC/PAL) later). Freeze frame on moving dot using interlaced scan. (b) A progressive scan. Freeze frame on moving dot using progressive scan.

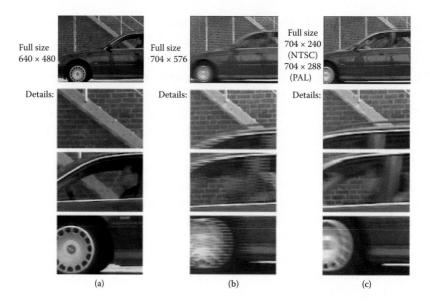

Figure 4.26 Comparison among (a) progressive, (b) interlaced, and (c) 2CIF-based scanning techniques. In these examples, the cameras were using the same lens. The speed of the car was the same in all cases, at 20 kilometers per hour (15 miles per hour). (a): Used in network cameras, full size 640×480; (b) used in analog CCTV cameras, full size 704×576; and (c) used in DVRs, full size 704×240 (PAL).

4.4.3 Progressive Scanning

With a progressive scan image sensor, values are obtained for each pixel on the sensor and each line of image data is scanned sequentially, producing a full-frame image. That is, captured images are not split into separate fields as in interlaced scanning. With progressive scan, an entire image frame is sent over a network and when displayed on a progressive computer monitor, each line of an image is put on the screen one at a time in perfect order, refreshing every 30th (NTSC) or 25th (PAL) of a second. There is virtually no "flickering" effect. Moving objects are therefore better presented on computer screens using the progressive scan technique. In a video surveillance application, it can be critical in viewing details of a moving subject (e.g., a person running away).

When a camera captures a moving object, the sharpness of a still image will depend on the technology used. Compare the JPEG images in Figure 4.26, captured by three different cameras using progressive scan, 4CIF interlaced scan, and 2CIF, respectively. See Section 4.6 for more information on resolution.

Note the following in the images in Figure 4.26:

- All image systems produce a clear image of the background.
- With interlaced scan images, jagged edges result from motion.
- In the 2CIF sample, motion blur is caused by lack of resolution.
- Only progressive scan makes it possible to identify the driver.

4.5 Image Processing

The lens and sensor are the key components in a camera that determine much of the quality of an image. The quality can be improved further by a network camera's image processor, which can adjust or apply various techniques and parameters that include:

- Control of exposure time, iris, and gain
- Backlight compensation and wide dynamic range
- Bayer demosaicing: converting raw data into a color image
- Noise reduction
- Color processing (e.g., white balance)
- Image enhancement (e.g., sharpening and contrast)

4.5.1 Exposure

The human eye can adapt to different lighting conditions, enabling us to see in very dark environments and extremely bright environments. A camera also must be able to cope with changes in brightness. To do this, the camera must find the correct exposure.

In a camera, there are three basic parameters used to achieve the appropriate exposure and therefore the ideal image quality:

1. Exposure time (how long the sensor is exposed to incoming light)
2. Iris or F-stop (how much light is coming through the lens to the sensor)
3. Gain (the amplification of the image level to make the images look brighter)

Exposure time, iris, and gain are often set automatically by the camera, but many cameras also have direct or indirect means of setting them manually. By making adjustments to one or more of these factors, an image of a scene will appear relatively unchanged even when the lighting at the scene changes.

The two images in Figure 4.27 are taken under different illumination conditions. Both images appear relatively similar, thanks to the auto exposure.

It is important to be aware that increasing the gain of the image not only increases the image luminance but also increases the noise. Adjusting the exposure time or opening the iris is therefore preferred. However, sometimes the iris is fully open and the exposure time required is longer than the time between two images (e.g., 1/25 second). The question then becomes: should the frame rate (i.e., the number of frames delivered every second) decrease or should the gain be increased? The answer depends on the application requirements and what the priorities are.

4.5.2 Backlight Compensation

Although a camera's automatic exposure tries to get the brightness of an image to appear as the human eye would see a scene, it can be easily fooled. Strong backlight is a difficult scenario for a camera to handle (Figure 4.28).

In the case with strong backlight, such as in the image in Figure 4.28 (top), the camera believes that the scene is very bright, and hence the camera reduces the iris opening or the exposure time, which results in a dark image. To avoid this, a mechanism called backlight compensation is introduced. It strives to ignore limited areas of high illumination, just as if they were not present. The resulting image (Figure 4.28, bottom) enables visual observation of all the objects in the scene. The bright areas, however, are overexposed. Without backlight compensation, the image would be too dark, and color and details would be lost in the foreground. Such a situation also might be dealt with by increasing the dynamic range of the camera, as discussed in the next section.

In addition to dealing with limited areas of high illumination, a network camera's automatic exposure must decide what area of an image should determine the exposure value. For example, the foreground (usually the bottom section of an image) can hold more important information than the background, for example, the sky (usually the top section of an image). The less important areas of a scene should not determine the overall exposure. The solution for the camera designer is to divide the image into sub-images and assign different weights for the exposure algorithm (Figure 4.29). In advanced network cameras, the user is able to select which of the predefined areas should be more correctly exposed. The position of the window or exposure area can be set to center, left, right, top, or bottom.

Figure 4.27 (Top): The top of the desk at center has an illuminance of 500 lux. (Bottom): The top of the desk at center has an illuminance of 30 lux.

Figure 4.28 (Top): Strong backlight, without backlight compensation. (Bottom): With backlight compensation applied, limited areas of high illumination can be ignored.

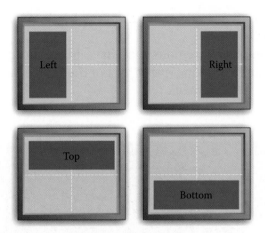

Figure 4.29 Assigning different weights to sections of an image to better determine which area should be more correctly exposed.

4.5.3 Wide Dynamic Range

Assigning different weights for different sections of an image is one way to determine exposure values. A problem occurs if there is a scene with areas that are both extremely bright (lit by sun) and dark (shaded areas) or backlight situations where a shadowed person is in front of a bright window.

In such cases, a typical camera would produce an image where objects in the dark area of the scene would hardly be visible. To solve this problem, some advanced cameras offer a feature called wide or high dynamic range (Figure 4.30), which incorporates techniques that handle a wide range of lighting conditions (i.e., scenes with large light contrasts or backlight situations). The true dynamic range of a scene is the range of light levels from the darkest object to the brightest object.

Techniques that enable high dynamic range typically use different exposures for different objects within a scene and employ ways to display the results, although a display screen also may have limited dynamic range.

Although high dynamic range may solve some problems, it also can introduce others. For example:

- Noise can be very different in different regions of an image; in particular, dark regions may contain very visible levels of noise.
- Pixels between two different exposure regions may show large visible artifacts. This can be seen in images with high

dynamic content and many different levels of lighting at the same time.

- Different exposure regions may have been allocated too low a dynamic range, making every part of the image bad.
- Colors may be very weak.

Figure 4.30 (Top) Image without wide dynamic range and (bottom) with wide dynamic range applied.

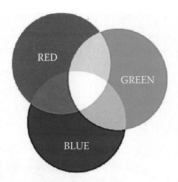

Figure 4.31 When red, green, and blue light are mixed in equal amounts, white is the result.

4.5.4 Bayer Demosaicing

After capturing a raw sensor image, it is ready for processing by the camera to produce a high-quality color image.

In a process called demosaicing, an algorithm is used to translate the raw data from the image sensor into full-color information at each pixel. Because each pixel only records the illumination behind one of the color filters, it needs the values from the other filters — interpolated from neighboring pixels — to calculate the actual color of a pixel. For example, if a pixel with a red filter registers a bright value and the neighboring green and blue pixels are equally bright, then the camera's processor determines that the color of the pixel with the red filter actually must be white (Figure 4.31).

4.5.5 Noise

Because no sensor is perfect, all cameras have an uncertainty with regard to the pixel values produced. This is known as random noise. It means that the same pixel will not give the exact same value each time it is read out, even if the scene and the illumination are the same. Sometimes, random noise appears as banding when entire rows are affected. Every sensor also has a few bad, nonworking pixels.

There are also differences between individual pixels within a typical sensor. This means that adjacent pixels exposed to exactly the same light generally will not respond exactly the same way. This will appear as a fixed-pattern noise that will not change over time. Some fixed-pattern noise will change with temperature and exposure time and, thus,

Figure 4.32 Half of the image above is shown with noise. Noise appears as specks of colors that distort the image.

will be more pronounced in hot environments or when exposure time is increased at night.

Part of the noise is generated within the camera, but part of it actually is due to the nature of light itself and thus affects all cameras. The latter is mostly visible in bright daylight images, for example, in a blue sky. Noise is lower in cameras that use a sensor with larger pixel sizes, which can receive more light in each pixel.

An image taken in low light might appear grainy or have specks of color (Figure 4.32) because the noise is amplified together with the low signal. This is true for both random noise and fixed-pattern noise. Fixed-pattern noise appears as if it is glued to the screen in a video camera but cannot be distinguished from random noise in a still-image camera.

Noise can be reduced by various filtering techniques that replace flawed pixels with new values calculated from neighboring pixels. Most cameras incorporate one or several filters in their processing. Most filters, however, have visible side effects that will appear as reduced resolution, motion blur, or other artifacts.

4.5.6 White Balance

Once color interpolation is done for an image, white balance is performed to ensure that the image has the right color balance (Figure 4.33).

To achieve the right color, neutral (black, gray, white) colors in a scene should stay neutral in the resulting image, regardless of the

(a)

(b)

Figure 4.33 Image with a reddish tint (a), with a greenish tint (b), and with white balance applied (c).   Continued.

(c)

Figure 4.33 Continued.

illumination. White balance can be set in a network camera by choosing either auto or one of any presets for indoor and outdoor settings. With auto white balance, the camera uses two or three different gain factors to amplify the red, green, and blue signals.

4.5.7 Sharpening and Contrast

Digital images can be fine-tuned using two tools: (1) digital sharpening and (2) contrast enhancement (Figure 4.34). Digital sharpening has to do with edge contrast, not resolution. Sharpening increases the local contrast by lightening the light pixels and darkening the dark pixels at the edges. Note that sharpening can also amplify noise. Contrast enhancement, on the other hand, affects not only the edges, but all pixels in the image equally. Contrast enhancement is handled by changing how the original pixel values are mapped or converted to values to be used by a display screen.

4.5.8 Aliasing

In cases when a subject contains finer details than the size of the pixels in an image sensor, the sensor cannot detect such details because the pixel

(a)

Figure 4.34 Sharpening versus contrast. (a): with no sharpening or changes in contrast levels; (b): contrast reduced; the darker areas become more visible and the lighter areas become less bright; (c): with sharpening applied. Continued.

resolution is too low. To see fine details, the image scale must be increased with a telephoto lens, or a sensor with a higher resolution must be used.

A problem occurs if a sensor views a repeated, unresolved pattern. The unresolved pattern then will cause distortions in the form of unwanted, larger patterns that can even be colored in the image. This phenomenon is known as aliasing. Fine patterns (e.g., the herringbone pattern on tweed jackets) may cause this effect. Patterns leading to aliasing, however, are rare. Some cameras have anti-aliasing filters to avoid this effect.

Repeated patterns can cause problems for some cameras due to an effect called aliasing. The effect is due to poor resolution. The two images in Figure 4.35 are representations of the same pattern on a jacket as seen with two different image sensor resolutions; the image at right has a pronounced aliasing effect.

Cameras translate continuous grades of tone and colors to points on a regular sampling grid. When details are finer than the sampling

(b)

(c)

Figure 4.34 Continued.

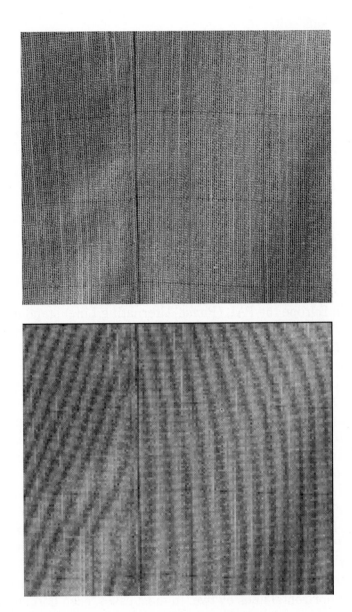

Figure 4.35 Two images that are representations of the same pattern on a jacket as seen with two different image sensor resolutions. Note pronounced aliasing effect in image on the bottom.

frequency, a good camera "averages" the details to avoid aliasing. This discards irresolvable details, but if too many details are filtered out, the image may look soft.

4.6 Resolution

Resolution in an analog or digital world is similar, but there are some important differences in how it is defined. In analog video, an image consists of lines, or TV-lines, because analog video technology is derived from the television industry. In a digital system, an image consists of square pixels (short for picture elements).

4.6.1 NTSC and PAL Resolutions

In North America and Japan, the NTSC (National Television System Committee) standard is the predominant analog video standard, whereas in Europe the PAL (Phase Alternating Line) standard is used. Both standards originate in the television industry. NTSC has a resolution of 480 lines and uses a refresh rate of 60 interlaced fields per second (or 30 full frames per second). A new naming convention, which defines the number of lines, type of scan, and refresh rate, for this standard is 480i60 ("i" stands for interlaced scanning). PAL has a resolution with 576 lines and uses a refresh rate of 50 interlaced fields per second (or 25 full frames per second). The new naming convention for this standard is 576i50. The total amount of information per second is the same in both standards.

When analog video is digitized, the maximum number of pixels that can be created is based on the number of TV lines available for digitization. In NTSC, the maximum size of the digitized image is 720×480 pixels. In PAL, the size is 720×576 pixels (D1). The most commonly used resolution is 4CIF: 704×576 (PAL) or 704×480 (NTSC). 2CIF resolution is 704×240 (NTSC) or 704x288 (PAL) pixels, which means dividing the number of horizontal lines by two. In most cases, each horizontal line is shown twice — so-called "line doubling" — when shown on a monitor, in an effort to maintain correct ratios in the image. This is a way to cope with motion artifacts because of interlaced scan. Section 4.4.2 discusses deinterlacing, and Figure 4.36 depicts the different NTSC and PAL image resolutions.

Sometimes, a quarter of a CIF image is used, called QCIF, short for Quarter CIF.

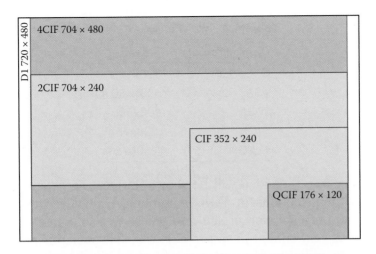

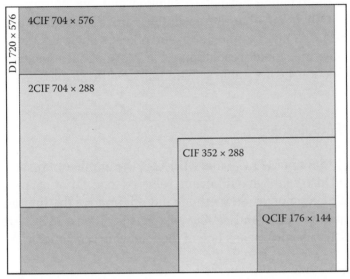

Figure 4.36 (Top) Different NTSC image resolutions; and (bottom) different PAL image resolutions.

4.6.2 VGA Resolutions

With the introduction of network cameras, 100 percent digital systems can be designed. This renders the limitations of NTSC and PAL irrelevant. Several new resolutions derived from the computer industry have been introduced. They are worldwide standards and provide better flexibility.

VGA is an abbreviation for Video Graphics Array, a graphics display system for PCs originally developed by IBM. The resolution is defined as 640×480 pixels, which is a common format used by non-megapixel network cameras. The VGA resolution is normally better suited for network

Table 4.4 VGA Resolution

Display Format	Pixels
QVGA (SIF)	320×240
VGA	640×480
SVGA	800×600
XVGA	1024×768
4x VGA	1280×960

cameras because the video will be shown, in most cases, on computer screens with resolutions in VGA or multiples of VGA. Quarter VGA (QVGA), with a resolution of 320×240 pixels, is also a commonly used format and is very similar in size to CIF. QVGA is sometimes called SIF (Standard Interchange Format) resolution and can be easily confused with CIF. Other VGA-based resolutions are SVGA (800×600 pixels), XVGA (1024×768 pixels), and 1280×960 pixels, which is four times VGA and provides megapixel resolution. (For more about megapixel resolution, see Section 4.6.4.). Table 4.4 summarizes the formats.

4.6.3 MPEG Resolution

In the early days of MPEG-1 and MPEG-2, the number of resolutions was limited and also somewhat different, referred to as D1 or parts of D1. The actual MPEG compression methods did not impose the limitations, just implementation limits and *de facto* standards. Today, MPEG resolution (Figure 4.37) usually means one of the following resolutions:

- 704×576 pixels (PAL 4CIF)
- 704×480 pixels (NTSC 4CIF)
- 720×576 pixels (PAL or D1)
- 720×480 pixels (NTSC or D1)

4.6.4 Megapixel Resolution

A network camera that offers megapixel resolution uses a megapixel sensor to deliver an image that contains one million or more pixels. The more pixels a sensor has, the greater the potential it has for capturing finer details and for producing a higher-quality image. Megapixel network cameras can be used to allow users to see more details (ideal for identification

Figure 4.37 Different MPEG image resolutions.

of people and objects) or to view a larger area of a scene. This benefit is an important consideration in video surveillance applications.

Table 4.5 summarizes some megapixel formats.

In the video surveillance industry, some best practices have emerged regarding the number of pixels required for certain applications. For an overview image, it generally is considered that about 70 to 100 pixels are enough to represent one meter (20 to 30 pixels per foot) of a scene. For applications that require detailed images, such as face identification, the demands can rise to as many as 500 pixels per meter (150 pixels per foot). This means, for example, that if it is necessary to identify people passing through an area that is 2 meters wide and 2 meters high, the camera must be able to provide a resolution of at least 1 megapixel (1000×1000 pixels).

Table 4.5 Megapixel Formats

Display Format	No. of Megapixels	Pixels
SXGA	1.3	1280×1024
SXGA+ (EXGA)	1.4	1400×1050
UXGA	1.9	1600×1200
WUXGA	2.3	1920×1200
QXGA	3.1	2048×1536
WQXGA	4.1	2560×1600
QSXGA	5.2	2560×2048

Figure 4.38 Megapixel camera view versus analog camera view.

Megapixel resolution is one area in which network cameras excel over analog cameras (Figure 4.38). The maximum resolution a conventional analog camera can provide after digitizing the video signal in a digital video recorder (DVR) or a video encoder is D1, which is 720×480 pixels (NTSC) or 720×576 pixels (PAL). The D1 resolution corresponds to a maximum of 414,720 pixels, or 0.4 megapixel. By comparison, a common megapixel format is 1280×1024 pixels, giving a 1.3-megapixel resolution, which is more than three times higher than the resolution that analog CCTV cameras can provide. Network cameras with 2-megapixel and 3-megapixel resolutions are also available, and even higher resolutions are expected in the future.

4.6.5 Aspect Ratios

With megapixel resolution, there also is a greater degree of flexibility in terms of the different aspect ratios in which images can be delivered (Figure 4.39). (Aspect ratio is the ratio of the width of an image to its height.) A conventional TV monitor displays an image with an aspect ratio of 4:3. Network video can offer the same ratio, in addition to others (e.g., 16:9). The advantage of a 16:9 aspect ratio is that unimportant details — usually located in the upper and lower part of a conventional-sized image — are not present and, therefore, bandwidth and storage requirements can be reduced.

Figure 4.39 Different aspect ratios.

4.6.6 High-Definition Television (HDTV) Resolutions

HDTV is a digital standard that defines that if a video display is capable of providing a minimum of 720p (progressive) or 1080i (interlaced) active video lines in a 4:3 aspect ratio and at least 540p/810i in a 16:9 aspect ratio, it is an HDTV display. Video in HDTV resolution contains much more detail because the number of pixels is larger. Compatibility with computer displays is better than traditional CCTV because the same square pixel format is used. No pixels need to be stretched for optimal viewing. With progressive scan HDTV video, no conversion (such as deinterlacing) is needed when the video will be processed by a computer or displayed on a computer monitor.

Table 4.5 and Table 4.6 list the basic HDTV image sizes in the European Broadcasting Union (EBU) and NTSC countries, respectively.

Table 4.5 Basic HDTV Image Sizes in the EBU

Size	Aspect Ratio	Scan	Frame Rate (fps/Hz)	Label
1280×720	16:9	Progressive	50	720p50
1920×1080	16:9	Interlaced	25[a]	1080i50
1920×1080	16:9	Progressive	25	1080p25
1920×1080	16:9	Progressive	50	1080p50

[a] 50-Hz field rate. Note that other frame rates can be used; the most common are 24, 25, 30, 50, 60 fps.

Table 4.6   Basic HDTV Image Sizes in NTSC Countries

Size	Aspect Ratio	Scan	Frame Rate (fps/Hz)	Label
1280×720	16:9	Progressive	60	720p60
1920×1080	16:9	Interlaced	30[a]	1080i30
1920×1080	16:9	Progressive	30	1080p30
1920×1080	16:9	Progressive	60	1080p60

[a] 60-Hz field rate. Note that other frame rates can be used; the most common are 24, 25, 30, 50, 60 fps.

Chapter 5

Video Compression Technologies

Analog video contains a tremendous amount of information, and when digitized it can consume as much as 165 Mbps (Mbps = mega bits per second) of bandwidth. That amount of data would not be practical to transmit over an IP (Internet Protocol)-based network or cost effective to store.

To address this, image and video compression techniques are utilized to reduce the bit rate. The goal is to reduce the amount of data as much as possible but, at the same time, have as little an impact on the image and video quality as possible. Depending on the application, different compression formats might apply. Often, the compression formats are referred to as "codecs." Codec is the abbreviated form of "**co**mpressor-**dec**ompressor" or "**co**ding-**dec**oding."

This chapter discusses the basics of compression as well as the most common compression formats and the standardization organizations behind them. Detailed information about the MPEG-4 compression and some best practices are provided at the end.

5.1 The Basics of Compression

Compression involves reducing image data by removing redundant information. As discussed previously, a digitized analog video sequence can comprise up to 165 Mbps of data. The data may have large amounts of redundant information, as in the case of an image with a white wall or blue sky.

All compression techniques are based on an understanding of how the human brain and eyes work together to form a complex visual system. Although an image may be optimally compressed, if it is not viewed in the original size or rate, artifacts may be visible. If pausing or zooming in on a recorded image in a video sequence is required, the recorded video should have a slightly larger bit rate (size) than that required for live viewing.

Listed below are some of the techniques commonly employed to reduce the size of images and video sequences:

- Reducing color nuances within an image (quantization)
- Reducing color resolution (subsampling)
- Removing small invisible parts (transform coding followed by quantization)
- Predicting parts of an image based on adjacent parts in the same image (intra-prediction)
- Removing repeated pixel values (run-length-coding or prediction)
- Efficient coding of pixels (entropy coding; for example, Huffman coding)
- Comparing adjacent images and removing unchanged detail (inter-prediction)

Some techniques are image-based compression techniques — also called intra-frame compression — where only one frame is evaluated and compressed at a time. Others are video compression techniques, or inter-frame compression, where several adjacent frames are compared as a way to further reduce the image data.

5.1.1 Image and Video Compression

Consider the video sequence displayed in Figure 5.1. The sequence shows a man running from right to left and a house that is stationary. With image compression techniques, each image in the sequence is coded as a separate unique image.

When the same video sequence is encoded using video compression techniques such as MPEG, the stationary part of the image is included only in the first frame. In the following two frames, only the moving parts (i.e., the running man) are included (Figure 5.2). Because an encoded sequence that uses a video compression technique contains less information, less

Figure 5.1 A three-image sequence, with a man running toward a house.

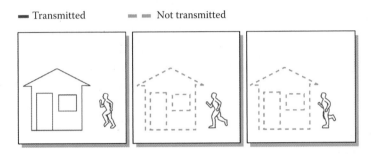

Figure 5.2 A three-image MPEG video sequence.

bandwidth and storage are required. When such an encoded sequence is displayed, the images appear as in the original video sequence.

5.1.2 Lossless and Lossy Compression

There are two basic categories of compression: (1) lossless and (2) lossy. Lossless compression is a class of techniques that allows for reconstruction of the exact original data from the compressed data. A limited number of techniques are made available to reduce the data, and, as a result, data reduction is limited. GIF (Graphics Interchange Format) is an example of a lossless image compression technique. Because of its limited compression abilities, it is not relevant for video surveillance.

Lossy compression, on the other hand, means that the compression data is reduced to an extent that the original information cannot be obtained when the video is decompressed. The difference is called artifacts.

5.1.3 Block and Transform

The compression algorithm can be pixel based, line based, or block based with selectable block size. A block is based on a number of pixels; for example, an 8 × 8 block is a block of 64 pixels, which is used in JPEG (named for the Joint Photographic Experts Group). The block size is the smallest complete part that algorithms can use to make calculations. Many popular algorithms use a mathematical block transform to sort the data before compression.

5.1.4 Prediction

Another image and video compression technique is prediction, which means predicting the next pixels based on the block of pixels nearby. If done well, an image with a blue sky can be coded by transferring one blue pixel and instructions to use the same until the sky is complete.

The prediction also can extend to predict the next image once the first image is sent. Image-to-image prediction is used in a video compression algorithm but not in an image compression algorithm. This is the basic difference between the two types of algorithms.

5.1.5 Latency

When compression takes place, one or several mathematical algorithms are used to remove image data. Similarly, to view a video file, algorithms are applied to interpret and display the data on a monitor. This process requires a certain amount of time, and the resulting delay is known as compression latency. The more advanced the compression algorithm, the higher the latency, given the same processing power.

Today, with increasing computing power available in network cameras and PC servers, latency is, in reality, not a big problem. The latency of a network, however, must be considered when designing a network video system.

For some applications, such as the compression of studio movies, compression latency is irrelevant because the video is not viewed live. In video surveillance applications where live monitoring takes place — especially when PTZ (pan, tilt, zoom) cameras are used — low latency is essential.

5.1.6 Compression Ratio

A compression ratio is defined as the ratio of the resulting file size compared with the original. A 50 percent compression ratio means that 50 percent of the original information has been removed and the resulting file is only half the size of the original file.

With efficient compression techniques, a significant reduction in file size can be achieved with little or no adverse effect on the visual quality. The extent to which the image modifications are perceptible depends on the degree to which the chosen compression technique is used. Often, a 50 to 95 percent compression is achievable with no visible difference, and in some scenarios a more than 98% compression ratio is possible.

5.2 Compression Standards

Standards are important in ensuring compatibility and interoperability. They are particularly relevant to video compression because video can be used for different purposes and, in some video surveillance applications, must be viewable many years from the recording date.

In the mid-1990s when storage was relatively expensive and standards for digital video compression were in their infancy, many video surveillance manufacturers developed proprietary video compression techniques. Today, most vendors use standardized compression techniques because they are as good as or better than the proprietary techniques. By deploying standards, end users are able to pick and choose from different vendors rather than be tied to one supplier when designing their video surveillance systems.

5.2.1 The ITU and ISO

There are two important organizations that develop image and video compression standards: (1) the International Telecommunication Union (ITU) and (2) the International Organization for Standardization (ISO).

The ITU is not a formal standardization organization. The ITU releases its documents as recommendations, for example, "ITU-R Recommendation BT.601" for digital video. The ISO is a formal standardization organization that cooperates with the International Electrotechnical Commission (IEC) for standards within areas such as IT. The latter organizations often are referred to as a single body, the "ISO/IEC."

The fundamental differences are the ITU stems from the telecommunications world and has standards relating to telecommunications, the ISO is a general standardization organization, and the IEC is a standardization organization dealing with electronic and electrical standards. However, given the ongoing convergence of communications and media, the organizations and their members have experienced an increasing overlap in their standardization efforts.

5.2.2 History of Compression Formats

The two basic compression standards are (1) JPEG (named for the Joint Photographic Experts Group) and (2) MPEG (named for the Moving Picture Experts Group), which are international standards set by the ISO and IEC, with contributors from the United States, Europe, and Japan, among others. JPEG and MPEG also are recommendations proposed by the ITU, which has further helped to establish them as global standards for digital still image and video coding. Within the ITU, the Video Coding Experts Group (VCEG) is the subgroup that has developed, for example, the H.261 and H.263 recommendations for videoconferencing over telephone lines.

The creation of the JPEG and MPEG standards began in the mid-1980s when the JPEG was formed. Seeking to develop a standard for color image compression, the group's first public contribution was the release of the first part of the JPEG standard in 1991. Since then, the JPEG group has continued to work on both the original JPEG standard and the JPEG 2000 standard, which never became popular.

In the late 1980s, the MPEG was formed with the purpose of deriving a standard for the coding of moving pictures and audio. It has since produced the MPEG-1, MPEG-2, and MPEG-4 standards.

At the end of the 1990s, a new group, the Joint Video Team (JVT), formed; it consisted of both the VCEG and MPEG. The purpose was to define a standard for the next generation of video coding. When this work was completed in May 2003, the result was MPEG-4 AVC/H.264. It was simultaneously launched as a recommendation by ITU ("ITU-T Recommendation H.264, Advanced video coding for generic audiovisual services") and as a standard by the ISO/IEC ("ISO/IEC 14496-10 Advanced Video Coding").

Sometimes the term "MPEG-4 Part 10" is used. This refers to the fact that the ISO/IEC standard that is MPEG-4 actually consists of parts, the current one being MPEG-4 Part 2. The new standard developed by the JVT was added to MPEG-4 as a separate part — Part 10 — called "Advanced Video Coding." This also is where the commonly used abbreviation AVC derives.

5.3 Compression Formats

This section provides a description of the compression formats that are, or have been, relevant to video surveillance.

Figure 5.3 Original image (left) and JPEG compressed image (right) using a high compression ratio that results in blockiness.

5.3.1 JPEG

The JPEG standard, ISO/IEC 10918, is the most widely used image compression format today. It is the most common format used in digital snapshot cameras and is supported by all Web browsers, which makes it widely accepted and easy to use. Users have the flexibility of having a high image quality with a fairly high compression ratio, or a very high compression ratio with lower image quality. Systems such as cameras and viewers can be made inexpensively due to the low complexity of the technique.

There is normally no visual difference between a JPEG compressed image and the original uncompressed image. However, if the compression ratio is pushed too high, artifacts in the form of "blockiness" appear (Figure 5.3).

The JPEG compression standard contains a series of advanced techniques. The main technique responsible for actually compressing an image is the discrete cosine transform (DCT) followed by a quantization that removes the redundant information (the "invisible" parts). The compression level normally can be set as a percentage, from 1 percent to 99 percent, with 99 percent being the highest compression level and the smallest file size, but also with the most artifacts (Figure 5.4a through 5.4f).

5.3.2 Motion JPEG

Motion JPEG (M-JPEG) is a digital video sequence represented as a series of JPEG images. The advantages are the same as with single still JPEG images — flexibility both in terms of quality and compression ratio. In addition, because there is no dependency between the frames, it is

(a)

(b)

Figure 5.4 (a) In a typical network camera, the compression ratio can be set between 0 and 99 percent; (b) a JPEG image with no compression applied — frame size 71 KB; (c) a JPEG image with 50 percent compression applied — frame size 24 KB; (d) a JPEG image with 90 percent compression applied — frame size 14 KB; (e) a JPEG image with 95 percent compression applied — frame size 11 KB. Depending on the level of detail and complexity in a scene, different compression ratios can be used. In most cases, 50 percent or more can be used; in some, 90 percent or even 95 percent can be applied with no or little visible impact. Continued.

Figure 5.4 Continued.

2168 kbit/s 25.91 f/s

(e)

Figure 5.4 Continued.

robust, meaning that if one frame is dropped during transmission, the rest of the video will be unaffected.

The main disadvantage of Motion JPEG is that because it uses only a series of still images, it makes no use of any video compression techniques. The result is a lower compression ratio for video sequences than would be the case if video compression techniques such as MPEG were used. Motion JPEG is popular in applications where individual frames in a video sequence are required (e.g., for analysis) and where lower frame rates, typically 5 fps or lower, are used.

5.3.3 JPEG 2000

JPEG 2000 was created as the follow-up to the successful JPEG compression, with better compression ratios. The basis was to incorporate new advances in image compression research into an international standard. Instead of using the DCT technique, JPEG 2000 or ISO/IEC 15444 uses the wavelet transform algorithm.

The advantage of JPEG 2000 is that the blockiness of JPEG is replaced with a more overall fuzzy image, as observed in the right-hand image in Figure 5.5.

Whether the fuzziness of JPEG 2000 is preferable over the "blockiness" of JPEG is a matter of personal preference. However, JPEG 2000 enables higher compression ratios than JPEG for the same image quality. For moderate compression ratios, JPEG 2000 produces images typically

Figure 5.5 Original image (left) and JPEG 2000 compressed image (right) using a high compression ratio to emphasize the artifacts.

about 75 percent the size of JPEG at equal image quality. JPEG 2000 is, however, a far more complex compression technique.

Despite its advantage over JPEG, JPEG 2000 never became popular in video surveillance applications and is still not widely supported in Web browsers.

5.3.4 Motion JPEG 2000

As with JPEG and Motion JPEG, Motion JPEG 2000 also can be used to represent a video sequence. The advantage is the same as with JPEG 2000, that is, a slightly better compression ratio compared with JPEG but with greater complexity.

The disadvantage is similar to that of Motion JPEG. Because it is a still-image compression technique, Motion JPEG 2000 lacks the advantages of a video compression technique. Therefore, the compression ratios are lower than with video compression techniques.

An undesirable effect with Motion JPEG 2000 video is that artifacts in the images tend to "float around" between each frame. This is in contrast with Motion JPEG, where the artifacts remain in the same place for each frame in a video stream and thus look stable over time. The viewing experience with Motion JPEG 2000 is therefore not as good as with a Motion JPEG stream. Motion JPEG 2000 has never succeeded as a video compression technique.

5.3.5 H.261 and H.263

H.261 and H.263 are not international standards but are recommendations of the ITU. They both are based on the same technique as the

MPEG standards and can be seen as simplified versions of MPEG video compression.

H.261 and H.263 originally were designed for videoconferencing over telephone lines (i.e., low bandwidth). However, they lack some of the more advanced MPEG techniques for efficient bandwidth use.

H.261 and H.263 are not suitable for use in general digital video coding. They were, however, used in the early 2000s in some digital video recorders and encoders.

5.3.6 MPEG-1

The first public standard of the MPEG committee was MPEG-1, ISO/IEC 11172. Its first parts were released in 1993. MPEG-1 video compression is based on the same technique used in JPEG. In addition, it includes techniques for efficient coding of a video sequence.

MPEG-1 focuses on bitstreams of about 1.5 Mbps and was originally designed for storage of digital video on CDs. The focus is on the compression ratio rather than image quality. The video quality is similar to the quality produced by a traditional VCR (video cassette recorder) but in digital form.

5.3.7 MPEG-2

The MPEG-2 project focused on extending the compression technique of MPEG-1 to cover larger images and higher quality, at the expense of higher bandwidth usage. MPEG-2, ISO/IEC 13818 also provides more advanced techniques to enhance the video quality at the same bit rate. However, it requires more complex and expensive equipment. The primary intention of MPEG-2 was for storing movies onto DVDs.

5.3.8 MPEG-4

The next generation of MPEG is based on the same technique as MPEG-1 and MPEG-2. Again, the new standard is focused on new applications.

The most important features of MPEG-4, ISO/IEC 14496, are the support of low-bandwidth-consuming applications, such as mobile devices like cell phones, and applications requiring high-quality images and with virtually unlimited bandwidth. The MPEG-4 standard allows for any frame rate, whereas MPEG-2 was locked into 25 frames per second in PAL and 30 frames per second in NTSC.

When "MPEG-4" is cited in video surveillance applications today, it is usually MPEG-4 Part 2 that is referred to. This is the "classic" MPEG-4 video streaming standard, also called MPEG-4 Visual.

Some network video streaming systems specify support for "MPEG-4 short header," which is an H.263 video stream encapsulated with MPEG-4 video stream headers. MPEG-4 short header does not take advantage of any of the additional tools specified in the MPEG-4 standard and, as a result, the video quality is low.

5.3.9 H.264

MPEG-4 Part 10 AVC/H.264 (referred to simply as H.264 in the remainder of this chapter) is the latest MPEG standard for video encoding. The new standard addresses several weaknesses in previous MPEG standards. It provides good video quality at substantially lower bit rates than previous standards, with better error robustness or better video quality at an unchanged bit rate. The standard is further designed to give lower latency, as well as better quality for higher latency.

An additional goal of H.264 was to provide enough flexibility so that the standard could apply to a wide variety of applications with very different requirements on bit rate and latency. Indeed, a number of applications with different requirements have been identified for H.264:

- Entertainment video, including broadcast, satellite, cable, DVD (1 to 10 Mbps, high latency)
- Telecom services (below 1 Mbps, low latency)
- Streaming services (low bit rate, high latency)

Some movies are encoded with H.264, and typically new DVD players for high-definition DVD formats, such as HD-DVD and Blu-ray, support the format.

5.3.10 MPEG-7

MPEG-7 is a different kind of standard, as it is a multimedia content description standard and does not deal with the actual encoding of video and audio. MPEG-7 uses XML (eXtensible Markup Language) to store metadata, and it can be attached to a time code to tag particular events in a stream. Although MPEG-7 is independent of the actual encoding technique of the video, it is most commonly used with MPEG-4.

MPEG-7 is relevant for video surveillance, as it can be used, for example, to tag the contents and events of video streams for more intelligent processing in video management software or video analytics applications.

5.3.11 MPEG-21

MPEG-21 is a standard that defines the means of sharing digital rights, permissions, and restrictions for digital content. MPEG-21 is also an XML-based standard and is developed to counter illegitimate distribution of digital content.

5.3.12 Compression Format Timeline

Over the past ten years, most of the above compression standards have been used, or are still used, in the video surveillance industry. See the timeline chart in Figure 5.6.

5.4 More on MPEG Compression

MPEG-4 — and soon H.264 — is the compression standard of choice for more and more video surveillance applications. These fairly complex and

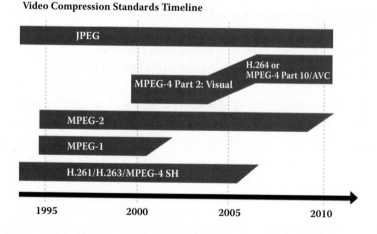

Figure 5.6 A timeline for the most common compression formats used in video surveillance. Eventually, JPEG and H.264 (MPEG-4 Part 10) will be the two dominant formats.

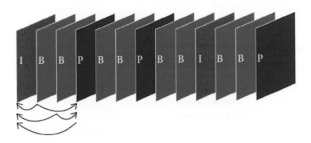

Figure 5.7 How a typical sequence with I-, B-, and P-frames might look. Note that a P-frame can reference only a preceding I- or P-frame, whereas a B-frame can reference both preceding and succeeding I- and P-frames.

comprehensive standards have some characteristics that are important to understand. The following subsections outline these characteristics.

5.4.1 Frame Types

The basic principle for video compression is the image-to-image prediction. The first image is called an I-frame and is self-contained, having no dependency outside that image. The following frames can use part of the first image as a reference (see Figure 5.7). An image predicted from one reference image is called a P-frame, and an image that is bi-directionally predicted from two reference images is called a B-frame.

- I-frames: intra-predicted, self-contained.
- P-frames: predicted from last I or P reference frame.
- B-frames: bi-directional; predicted from two references — one in the past and one in the future — and thus out-of-order decoding is needed.

The video decoder restores the video by decoding the bitstream frame by frame. Decoding must always start with an I-frame, which can be decoded independently, whereas P- and B-frames must be decoded together with the current reference image (or images).

5.4.2 Group of Pictures

One parameter that can be adjusted in MPEG-4 is the Group of Pictures (GOP) length and structure, also referred to as Group of Video (GOV)

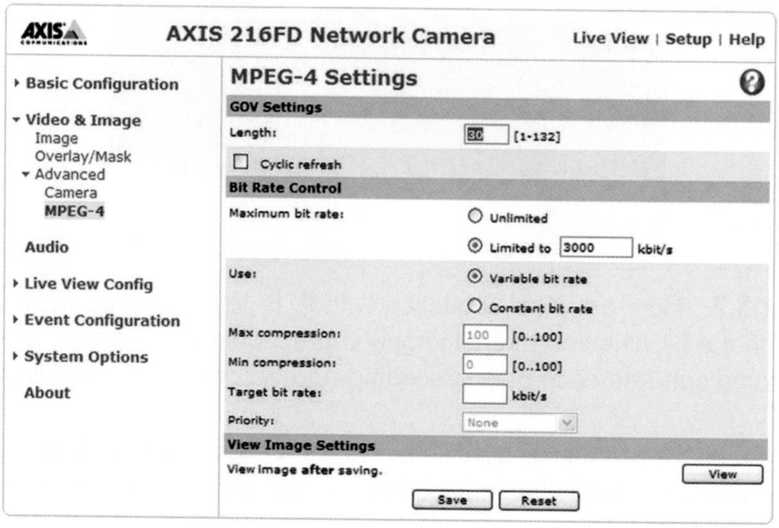

Figure 5.8 An interface in a network camera where the length of the Group of Video (GOV) (i.e., the number of frames between two I-frames) can be adjusted to fit the application.

in some MPEG standards (Figure 5.8). It is normally repeated in a fixed pattern, for example,

GOV = 4 (e.g., IPPP IPPP ...)
GOV = 15 (e.g., IPPPPPPPPPPPPPP IPPPPPPPPPPPPPP ...)
GOV = 8 (e.g., IBPBPBPB IBPBPBPB ...)

The appropriate GOV depends on the application. Decreasing the frequency of I-frames decreases the bit rate. Removing the B-frames reduces the latency.

5.4.3 Variable and Constant Bit Rates

Another important aspect of MPEG is the bit rate mode used. In most MPEG systems, it is possible to select the mode — constant bit rate (CBR) or variable bit rate (VBR) — to use. The optimal selection depends on the application and available network infrastructure.

With limited bandwidth available, the preferred mode is normally CBR because this mode generates a constant and predefined bit rate. The disadvantage with CBR is that image quality will vary. Whereas the quality will remain relatively high when there is no motion in a scene, it will significantly decrease with increased motion.

With VBR, a predefined level of image quality can be maintained, regardless of motion or the lack of it in a scene. This often is desirable in video surveillance applications where there is a need for high quality, particularly if there is motion in a scene. Because the bit rate in VBR can vary — even when an average target bit rate is defined — the network infrastructure (available bandwidth) for such a system needs to have a higher capacity.

5.4.4 Profile@Level

Because both MPEG-2 and MPEG-4 cover a wide range of image sizes, frame rates, and bandwidth usage, the MPEG-2 introduced a concept called Profile@Level. This was created to make it possible to communicate compatibilities among applications. For example, the Studio profile of MPEG-4 is not suitable for a PDA (personal digital assistant), and vice versa.

Examples of common profiles are MPEG-2 with its Main Profile at Main Level (MP@ML), MPEG-4 Main Profile at L3 Level, and H.264 Main Profile at Level 5.

The Internet Streaming Media Alliance (ISMA) is using the Profile@ Level definitions to ensure that devices streaming video over the Internet are compatible.

5.4.5 Licensing

MPEG-2, MPEG-4, and H.264 are subject to licensing fees. This requires any company manufacturing products using these compression standards to pay the appropriate licensing fees. MPEG LA, an independent license administrator, manages the licensing fees. For most network video products, one or several licenses have been paid by the manufacturer, which means that the video can be monitored on one or a few monitoring stations.

If an end user is planning to monitor the video at more monitoring stations than the product is licensed for, that end user must buy additional licenses to match the number of stations. If the manufacturer has not paid the license fees, it most likely means that the manufacturer does not follow the compression standard 100 percent, which in turn will limit the compatibility with other systems.

5.4.6 Backwards Compatibility

MPEG-2 and later standards are not backwards compatible; that is, strict MPEG-2 decoders and encoders will not work with MPEG-1. In addition,

H.264 encoders and decoders will not work with MPEG-2 or previous versions of MPEG-4 unless they are specifically designed to handle multiple formats. There are, however, various solutions available where streams encoded with newer standards can sometimes be packetized inside older standardization formats to work with older distribution systems.

5.4.7 Encoder Quality

It is important to note that all MPEG standards, including H.264, define the syntax of an encoded video stream, together with the method of decoding this bitstream. Thus, only the decoder is actually standardized. An MPEG encoder in a network camera or video encoder product can be implemented in different ways, and a vendor can choose to implement only a subset of the standard.

This helps to optimize the technology and reduce the complexity in implementations. However, it also means that MPEG video from different vendors can differ in quality. If video quality is a concern, it is always recommended that users buy and test a few products to ensure that the quality is appropriate for the application.

5.5 Best Practices

One compression standard and configuration does not fit all, and depending on the application, different compression standards and configurations might apply. When designing a network video application, consider the following questions:

- *Frame rate.* What is the required frame rate? Is the same frame rate required all the time? Below 5 frames per second, Motion JPEG is often the best choice and the frame rate can be controlled by video motion detection. For higher frame rates, MPEG-4 is normally the best because it saves bandwidth and storage.
- *Bandwidth.* What is the available network bandwidth? In scenarios with very low bandwidth, MPEG-4 compression using constant bit rate may be the only option, whereby image quality will be sacrificed when there is motion.
- *Image quality.* What is the allowed level of image degradation (artifacts) due to compression? Compression ratios well above 90 percent can be used if the scene is not too complex.

- *Latency.* What is the acceptable level of latency? If video is not monitored live but only recorded, latency might not be an issue. When controlling PTZ cameras, it is important to have low latency.
- *Robustness.* How robust or secure must the system be? Is it acceptable that video might be lost for 0.5 seconds if a frame is dropped on the network? What is the budget for the system?
- *Standard and compatibility.* Is the openness of the system and interoperability with other systems important? If so, make sure the chosen products follow the standard 100 percent.

Chapter **6**

Audio Technologies

Much of our learning and our relation with others is conducted through audiovisual means — that is, through what we see, hear, and say. In our daily lives, we often are alerted to unusual events first by what we hear. Then the events are visually verified.

Having audio as an integrated part of a video surveillance system can be an invaluable addition to a system's ability to, for example, detect and interpret events and emergency situations. To illustrate this, consider a video surveillance system without audio. A cry for help, the sound of breaking glass, a gunshot, or an explosion in the vicinity of a camera but outside the camera's field of view would escape notice by a video surveillance system without audio. Even if a scene such as a parking lot were under visual surveillance, without audio support the system would not be able to pick up, for example, a vehicle alarm. The ability of audio to cover a 360° area enables a video surveillance system to extend its coverage beyond a camera's field of view. It can instruct a PTZ (pan, tilt, zoom) or dome camera (or alert the operator of one) to visually verify an audio alarm. An audiovisual surveillance system therefore increases the effectiveness of a security solution by enabling a remote user to receive more information.

Audio also can be used in other ways. It can provide users with the ability to not only listen in on an area but also communicate orders or requests to visitors or intruders. For example, if a person in a camera's field of view demonstrates suspicious behavior — such as loitering near a bank machine — or is seen entering a restricted area, a remote security guard can send a verbal warning to the person. In a situation where a person has been injured, being able to remotely communicate with and

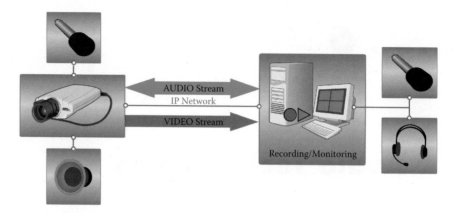

Figure 6.1 A network video system with integrated audio support. Audio and video streams are sent over the same network cable.

notify the victim that help is on the way also can be beneficial. Access control — that is, a remote "doorman" at an entrance — is another area of application. Other applications include a remote help desk situation (e.g., an unmanned parking garage), and videoconferencing.

Although having audio in a video surveillance system is still not widespread, its implementation is expected to increase as the adoption of network video increases. Network video enables easier implementation of audio than in an analog CCTV (closed-circuit television) system. In an analog system, separate audio and video cables must be installed from endpoint to endpoint, that is, from the camera and microphone location to the viewing or recording location. If the distance between the microphone and the station is too long, balanced audio equipment must be used, which increases installation costs and difficulty. In a network video system (Figure 6.1), a network camera with audio support processes the audio and sends both the audio and video over the same network cable for monitoring and recording. This eliminates the need for extra cabling and makes synchronizing the audio and video much easier.

Support for audio can be found in many types of network cameras, including fixed, PTZ, and dome cameras. Many camera manufacturers are recognizing the importance of audio, and audio is becoming a common feature in network cameras.

Some video encoders also have built-in support for audio, which means that they can provide audio functionality in an analog camera installation. This may be useful in an application where a specialty camera is used or if existing analog cameras are installed. Because the video encoder can be located close to the analog camera, the length of the audio cables can be reduced (Figure 6.2).

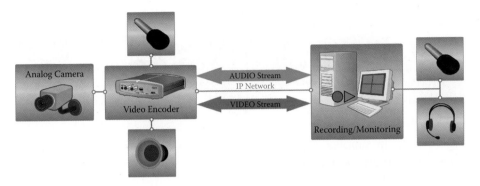

Figure 6.2 Some video encoders have built-in audio, making it possible to add audio even if analog cameras are used in an installation.

When audio is a consideration, its application should be clear because it affects what products should be selected. The following sections of this chapter discuss information on audio transmission modes (simplex, half duplex, or full duplex), audio equipment (microphones, speakers, and cabling), acoustical adjustments, audio detection alarms, codecs and bit rates, audio and video synchronization, and a summary of the factors that affect audio quality. For information about how the use of audio can sometimes be restricted or regulated by local legislation or codes of practice, see Chapter 15.

6.1 Audio Modes

Depending on the application, there might be a need to send audio in only one direction or both directions. This can be done either simultaneously or in one direction at a time. There are three basic modes of audio communication: simplex, half duplex, and full duplex.

6.1.1 Simplex

Simplex means audio can be sent in one direction only. Audio is sent either from the camera only — which is most often the case — or from the user only. Situations where audio is sent only from the camera include remote monitoring and video surveillance applications where live audio, as well as video, from a monitored site is sent over a network (Figure 6.3). Applications where audio is sent only from a user or operator include situations where there is a need to provide spoken instructions to a person seen on the camera or in a parking lot scenario where the operator can use audio to scare a potential car thief off the lot (Figure 6.4).

Figure 6.3 In simplex mode, audio is sent in one direction only. In this case, the camera sends audio to the operator.

Figure 6.4 In simplex mode, audio is sent in one direction only. In this case, the operator sends audio to the camera.

6.1.2 Half Duplex

Half duplex enables audio to be sent and received in both directions — from the camera and the operator — but only in one direction at a time (Figure 6.5). This type of communication is similar to a walkie-talkie. To speak, an operator must press and hold down a push-to-talk button. Releasing the button enables the operator to receive audio from the camera. Half-duplex communication has no echo problem (a topic discussed later in this chapter). This is the recommended mode if network bandwidth is limited.

Figure 6.5 In half-duplex mode, audio is sent in both directions, but only one party at a time can send.

Figure 6.6 In full-duplex mode, audio is sent to and from the operator simultaneously.

6.1.3 Full Duplex

Full duplex enables users to send and receive audio (talk and listen) at the same time (Figure 6.6). This mode of communication is similar to a telephone conversation. Full duplex requires that the client PC has a sound card with support for full-duplex audio.

6.2 Audio Equipment

When a network camera or a video encoder has support for audio, it may very well include a built-in microphone but very rarely a built-in speaker. The built-in microphone may be appropriate for some applications, although for many a better external microphone may provide a better solution. This section provides some guidance when selecting audio equipment.

6.2.1 Audio Input (Microphones)

A network camera with integrated audio functionality often provides a built-in microphone or a mic-in/line-in jack (Figure 6.7). A camera with mic-in/line-in support gives users the option of using another type or quality of microphone than the one built into the camera, and it also enables the camera to connect to more than one microphone. In addition, it enables a microphone to be located some distance away from the camera. For example, the camera can be located close to the ceiling while the microphone can be placed close to a door handle.

Mic-in enables an external microphone with no built-in amplifier to connect to a network camera that has a built-in amplifier. Line-in means the network camera can connect to devices that provide an already amplified audio signal (known as line signal). Devices that provide line signals

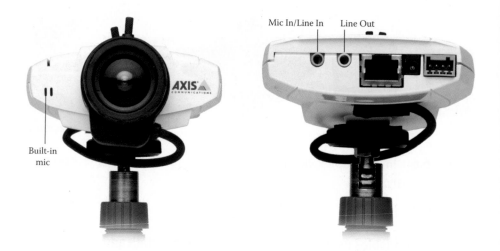

Figure 6.7 Front and back views of a network camera with audio support.

include mixers, which allow a network camera to connect to several microphones, and microphone amplifiers, which connect to microphones without built-in amplifiers.

There are three main types of microphones: (1) condenser, (2) electret condenser, and (3) dynamic. They differ in the way they convert sound into electrical signals.

6.2.1.1 Condenser Microphones

Of the three main types of microphones, the condenser offers the highest audio sensitivity and quality. It is often used in professional recording studios and can be used in video surveillance applications that require high audio quality. A condenser microphone is used together with a so-called phantom power supply, which supplies the necessary 48 volts required by the microphone. A condenser microphone uses an XLR connector. If a network camera does not support an XLR connector, a condenser microphone can still be connected to the camera via an adapter that may be provided by the phantom power supply box (Figures 6.8a and 6.8b).

6.2.1.2 Electret Condenser Microphones

The built-in microphone in a network camera is often an electret condenser microphone. This type of microphone is commonly used in headphones and computer microphones (Figure 6.9) and often uses a 3.5-mm audio connector. It offers a high level of sensitivity and is less expensive than a condenser microphone. The electret condenser microphone needs a voltage of around 2 v. If an external electret microphone is used with a

(a)

(b)

Figure 6.8 (a) A condenser microphone with a phantom power supply box; and (b) a condenser microphone for use on tabletops.

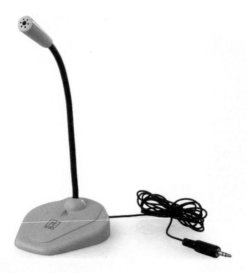

Figure 6.9 A computer microphone.

network camera, the network camera can supply the microphone with the necessary power.

6.2.1.3 Dynamic Microphones

The dynamic microphone is rarely used in the video surveillance industry because it does not have high audio sensitivity and has a poor ability to reproduce bass tones when the source of the sound is not close to the microphone. The dynamic microphone uses an XLR connector. If a network camera does not support an XLR jack, an adapter can be used to connect the camera with a dynamic microphone.

6.2.1.4 Directional Microphones

Microphones are made with different polar patterns (also called pick-up or directional patterns). The types of patterns include omnidirectional, which picks up audio equally in all directions, and unidirectional such as cardioid (meaning "heart-shaped") and supercardioid, which has high audio sensitivity in one specific direction. To pick up sounds in a specific spot located far away from a network camera, a specialized, unidirectional microphone called a shotgun microphone can be used.

6.2.2 Audio Output (Speakers)

There is a wide array of speakers available that can be used along with network cameras. PC speakers are used most often. The power of speakers

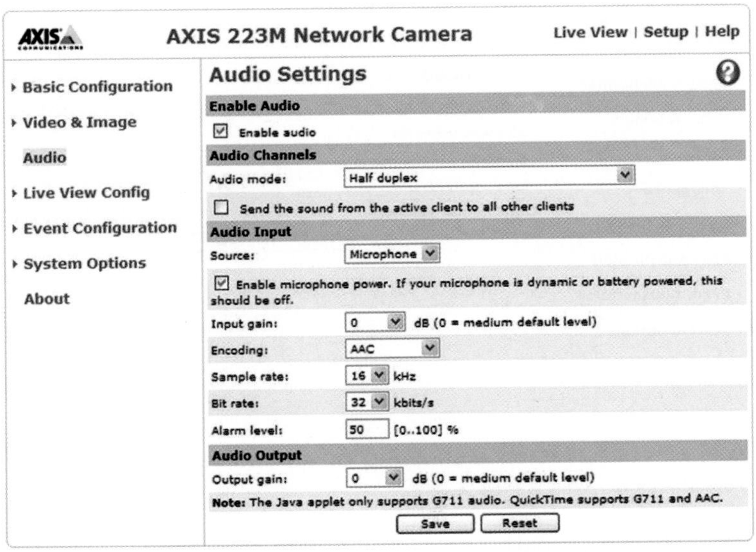

Figure 6.10 An example of a user interface for audio settings on a network camera.

is measured in watts. It indicates how much power the speakers consume and often relates to how loud the speakers can be.

An active speaker, which is a speaker with a built-in amplifier, can be connected directly to a network camera. If a speaker has no built-in amplifier, it must first connect to an amplifier, which is then connected to a network camera.

6.3 Acoustical Adjustments

There are a number of adjustments (Figure 6.10) that can be made to get the best audio quality from an installation. This section discusses several of the most common adjustments (i.e., volume and gain, echo cancellation, speech filter, and noise cancellation).

6.3.1 Volume and Gain

The sensitivity of a microphone can be adjusted if the connected microphone is an ordinary (unamplified) microphone, such as a computer microphone or clip-on microphone. To reduce noise, the gain setting on the camera (if using a camera's built-in microphone) or on the amplifier (if using a stand-alone microphone) should be adjusted so that the signal does not clip or distort. It is important to have a good signal gain

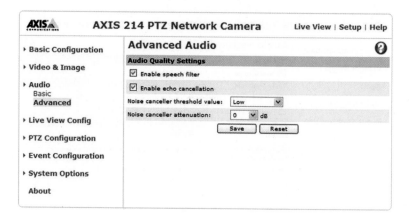

Figure 6.11 A user interface showing advanced settings for speech filter, echo cancellation, and noise canceller.

throughout the signal path and as much gain as possible early in the signal path (i.e., in the network camera).

6.3.2 Echo Cancellation

In full-duplex mode, the microphone can capture unwanted sounds from the speaker. Such sounds, known as feedback, can be reduced by facing the microphone away from the speaker and using echo cancellation (Figure 6.11). A network camera with echo cancellation has a memory chip that "remembers" the audio signals that have just been sent from a microphone for a short time. If the microphone picks up the same audio signals again shortly thereafter, the device recognizes the signals as echo from the speaker and will remove those particular signals.

6.3.3 Speech Filter

If voices are the only sounds of interest, a speech filter (technically known as a bandpass filter) (Figure 6.11) can be used to filter out low and high frequencies outside the voice frequency band, which ranges from approximately 250 to 3,500 Hz.

6.3.4 Noise Cancellation

Noise cancellation (Figure 6.11) can be used to reduce background noise. This feature is configured using two parameters: (1) threshold and (2) attenuation. Threshold is used to define the level under which noise will be reduced. Then attenuation can be used to choose the degree of noise reduction. Total background noise cancellation may not be desirable because a listener might interpret it as a break in the connection. Noise cancellation should be used cautiously because it can deteriorate the quality of the audio.

6.4 Audio Detection Alarm

In the same way that a network camera can analyze video, it can analyze audio. Audio detection alarm can be used as a complement to video motion detection because it can react to events in areas too dark for the video motion detection functionality to work properly. It can also be used to detect activity in areas outside the camera's view.

When sounds such as the breaking of a window or voices in a room are detected, they can trigger a network camera to send and record video and audio, send e-mail or other alerts, and activate external devices such as alarms. Similarly, alarm inputs such as motion detection and door contacts can be used to trigger video and audio recordings.

In a PTZ or network dome camera, audio alarm detection can trigger the camera to turn automatically to a preset location such as a specific window.

Audio detection can be enabled all the time, enabled during specific times, or disabled. It can be set to trigger an event if the incoming sound level rises above, falls below, or passes a certain level of sound intensity.

6.5 Audio Compression

Analog audio signals must be converted into digital audio through a sampling process and then compressed to reduce the size for efficient transmission and storage. The conversion and compression are done using an audio codec, an algorithm that codes and decodes audio data.

There are many different audio codecs supporting different sampling frequencies and levels of compression. Some audio codecs support either constant bit rate only mode or both constant and variable bit rates. All these factors affect the audio quality and file size.

6.5.1 Sampling Frequency

Sampling frequency or rate refers to the number of times per second a sample of an analog audio signal is taken and is defined in terms of hertz

(Hz). The human ear can hear sounds up to 20 kHz, and to capture this level of sound with good quality, a sampling frequency of at least 40 kHz is necessary. Music CDs, for example, use a sample rate of 44.1 kHz.

As a rule of thumb, the sample rate (according to the Nyquist–Shannon sampling theorem) must be twice the maximum required frequency. If the human voice is the only sound of interest, then a sample rate of at least 8 kHz is needed because the frequency of the human voice is normally below 4 kHz. In general, the higher the sampling frequency, the better the audio quality and the greater the bandwidth and storage needs are.

6.5.2 Bit Rate

The bit rate is an important setting in audio because it determines the level of compression and thereby the quality of the audio. In general, the higher the compression level (the lower the bit rate), the lower the audio quality. The differences in the audio quality of codecs can be particularly noticeable at high compression levels (low bit rates) but not at low compression levels (high bit rates). Higher compression levels also can introduce more latency or delay, but they enable greater savings on bandwidth and storage use.

The bit rates most often selected with codecs are between 32 and 64 kbps. Audio bit rates, as with video bit rates, are an important consideration when calculating total bandwidth and storage requirements.

Some codecs enable a constant bit rate mode only, whereas others enable both constant bit rate (CBR) and variable bit rate (VBR) modes. When using VBR, the bit rate adjusts to the complexity of the audio; that is, less demanding audio is compressed more and takes fewer bits than more complex sounds. This enables delivery of a higher-quality stream than a CBR file of the same size. When using VBR, a target bit rate can be set so that the levels of compression fluctuate close to the desired bit rate. The downside of using VBR is that the encoding time may be longer.

6.5.3 Codecs

Today there are three main audio codecs used in network video. They are AAC-LC, which requires a license, and G.711 and G.726, which are non-licensed technologies. Of the three codecs, only AAC-LC supports either constant or variable bit rate. The G.711 and G.726 standards support only constant bit rate. A fourth codec that is applicable for network video is G.722.2. More details on each of the four codecs follow.

6.5.3.1 AAC-LC

AAC-LC stands for Advanced Audio Coding — Low Complexity. The codec's official name is MPEG-4 AAC and includes four different profiles, where "LC" is the least complicated and therefore the most widely used form.

AAC offers sampling frequencies ranging from 8 to 96 kHz and bit rates ranging from as low as 2 kbps for low bit rate speech encoding to more than 300 kbps for high-quality audio codings. It supports constant and variable bit rate modes.

If achieving the best possible audio quality is a priority, AAC is the recommended codec to use, particularly at a sampling rate of 16 kHz or higher and at a bit rate of 64 kbps. If a product does not offer a sample rate of 16 kHz, then AAC with a sample rate of 8 kHz at a bit rate of 24 or 32 kbps is recommended. AAC-LC requires a license for encoding and decoding.

AAC-LC was developed by companies that include Dolby, Fraunhofer IIS, Sony, Nokia, and AT&T and has been part of the MPEG standard since 1997. It is specified as Part-7 of the MPEG-2 standard and Part 3 of the MPEG-4 standard. Because AAC-LC is part of the MPEG-4 standard, it is highly possible that it will become the most adopted standard in the video surveillance industry. For more information on the MPEG group, see Chapter 5 on compression technologies.

6.5.3.2 G.711 PCM

G.711 PCM (pulse-code modulation) is an unlicensed standard from the ITU-T (International Telecommunication Union — Telecommunication Standardization Sector). It was developed in 1972 as a telephony standard. All IP telephony and VoIP (Voice over IP) manufacturers follow this standard. It has a sampling frequency of 8 kHz and a bit rate of 64 kbps. This standard is very useful when integrating audio into a VoIP system.

6.5.3.3 G.726 ADPCM

G.726 ADPCM (adaptive differential pulse code modulation) is an unlicensed speech codec from the ITU-T. It has a sampling frequency of 8 kHz and bit rates of 16, 24, 32, and 40 kbps. The most commonly used bit rate is 32 kbps. G.726 is a low-power and low-cost implementation standard with low latency. It is the most commonly used codec within the security industry but is not widely used outside this industry. G.726 ADPCM was introduced in 1990.

6.5.3.4 G.722.2 or AMR-WB

G.722.2 or AMR-WB (adaptive multi-rate — wideband) is a licensed speech codec from the ITU-T. It has a sampling frequency of 16 kHz and offers bit rates ranging from 6.60 to 23.85 kbps. This is the codec used in networks such as UMTS, a 3G mobile phone technology. G.722.2 offers good speech performance at rates of 12.65 kbps and higher.

6.6 Audio and Video Synchronization

A media player (a computer software program used for playing back multimedia files) or a multimedia framework (such as Microsoft DirectX, which is a collection of application programming interfaces that handle multimedia files) handles the synchronization of audio and video data. Audio and video are two separate packet streams that are sent over a network. For the client or player to perfectly synchronize the audio and video streams, the audio and video packets must be time-stamped.

The time-stamping of video packets using Motion JPEG compression may not always be supported in a network camera. If this is the case and if having synchronized audio and video is important, it is better to use MPEG-4 or H.264 compression because such video streams are usually sent using RTP (Real-time Transport Protocol), which time-stamps the video packets.

There are, however, many situations where synchronized audio is less important or even undesirable — for example, if audio will be monitored but not recorded.

6.6 The Future of Audio in Network Video

Audio is likely to become a key consideration when selecting a video surveillance solution in the future. Audio, together with video, provides a more complete monitoring solution.

Future network cameras likely will support dual audio codecs simultaneously, just as two video compression formats (MPEG-4 and Motion JPEG) can be supported simultaneously. This will enable users to take advantage of the different strengths of the codecs and apply them for different purposes, for example, one codec for recording and another for communications only.

Other improved functionalities likely will lie in the following areas: synchronization between audio and video, audio analytics (real-time or post-event), audio tracking with PTZ cameras, audio quality, and the ability to search for certain sounds such as speech and breaking glass.

6.7 Best Practices

In video surveillance situations, a decent audio quality is normally acceptable. However, there are some things to consider when attempting to achieve the best quality from an installation.

- *Audio equipment and placement.* Choose and set up audio equipment based on application needs. The type and polar pattern of a microphone and the placement of the microphones, cabling, and speakers affect audio quality. Although an audio signal can be amplified later, good choice and placement of audio equipment will help reduce noise. The microphone should be placed as close as possible to the source of the sound. In full-duplex mode, a microphone should face away and be placed some distance from a speaker to reduce feedback from the speaker.
- *Amplify the signal as early as possible.* This minimizes noise in the signal chain. In addition, make sure the signal levels are as close to, but not over, the clipping level, which is the level when audio is distorted.
- *Acoustical adjustments.* Audio quality can be improved by adjusting the input gain and using different features such as echo cancellation and speech filter.
- *Choice of codec and bit rate selection.* The type of codec and bit rate level selected will affect audio quality. In general, the higher the compression level, the lower the audio quality.
- *Use shielded cable.* To minimize disturbance and noise, always use a shielded audio cable and avoid running the cable near power cables and cables carrying high-frequency switching signals. Also, keep audio cables as short as possible. If a long audio cable is required, balanced audio equipment — that is, cable, amplifier, and microphone that are all balanced — should be used to reduce noise.
- *Legal implications.* Some countries restrict the use of audio and video surveillance. It is a good idea to check with the local authorities.

Chapter 7

Video Encoders

Video encoders, also commonly called video servers, are key pieces of equipment that help analog CCTV (closed-circuit television) systems migrate into an open platform–based network video system. They will continue to play a significant role in the video surveillance market, as more than 95 percent of the estimated 40 million surveillance cameras installed worldwide are still analog.

Many of the analog cameras have been installed over the past few years and will continue to be functional for years to come. The average life expectancy of an analog surveillance camera is five to seven years. In many installations, the coaxial cable is the most expensive part of the installation; thus, once installed, there is often limited incentive to re-cable with Cat 5 to enable installation of a network camera. However, the recording device in an analog system, the VCR (video cassette recorder) or DVR (digital video recorder), usually fails long before the camera component. This is where the video encoder comes into play.

Video encoders allow security managers to retain analog CCTV cameras while building a video surveillance system that provides the benefits of network video. Using a video encoder, analog cameras can be controlled and accessed over an IP (Internet Protocol) network, such as a LAN (local area network) or the Internet, and old video recording equipment (DVRs or VCRs), as well as analog monitors, can be replaced with standard computer monitors and servers (Figure 7.1).

Analog cameras of all types, such as fixed, indoor, outdoor, fixed dome, dome, and PTZ (pan, tilt, zoom), as well as specialty cameras such as covert, miniature, and microscope cameras, can be integrated and controlled in a network video system using video encoders.

Figure 7.1 An analog camera connected to a video encoder, which makes it possible to include the camera in a network video system.

The following sections provide an overview of the components of a video encoder and the different types of video encoders available. Different types of video encoders can be used, depending on the configuration, number of cameras, camera type, and whether or not coax cabling is installed. Some best practices are outlined at the end of this chapter.

7.1 Components of a Video Encoder

A video encoder (Figure 7.2) connects to an analog camera via a coax cable and converts analog video signals into a compressed, digital video stream that is then transmitted over an IP-based network. It is called a

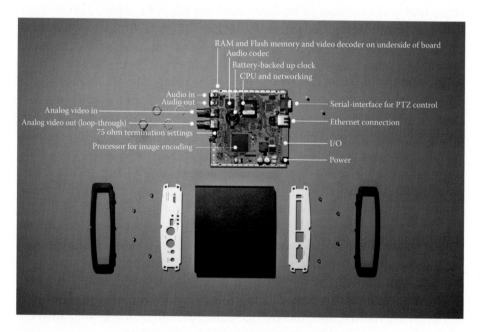

Figure 7.2 Components of a video encoder.

Figure 7.3 A video encoder with PoE functionality has the ability to power not only the video encoder itself but also the analog camera that is connected to it.

video encoder because it encodes video using a compression standard such as Motion JPEG (named for the Joint Photographic Experts Group) or MPEG-4 (named for the Moving Picture Experts Group). Once the video is on a network, it is identical to a video stream coming from a network camera and is ready to be integrated into a network video system. A video encoder also often includes a serial port, which is commonly used for controlling the PTZ functionality of the attached analog camera.

A video encoder also can provide many advanced functionalities, such as de-interlacing, video motion detection, alarm handling, one- or two-way audio support, and audio alarm.

Some video encoders also include Power over Ethernet (PoE) functionality (Figure 7.3). This enables both the video encoder *and* the analog camera connected to it to receive power through the same cable used for data transmission. This feature can provide substantial savings to the entire system because power cables can be excluded from an installation. In addition, if the server room is connected to an uninterruptible power supply (UPS), PoE enables the entire surveillance system, including the attached analog cameras, to receive centralized backup power and to operate even in the event of a power failure.

The networking functionality is also very important and should include all the latest security and IP protocols. The processor, which could be a general-purpose processor, a DSP (digital signal processor), or a purpose-built ASIC (application-specific integrated circuit), in the video encoder determines the performance, normally measured in frames per second per channel in the highest resolution, D1. Video encoders that can provide full frame rate (30 frames per second) in the highest resolution for all ports normally come at a premium, and there are only a few on the market with this level of performance.

Inside the encoder, there is also memory for storing the encoder's firmware (a computer program) and for local recording of video. In addition, most video encoders today have auto sensing, meaning that it will recognize if the incoming video signal is PAL (Phase Alternating

Line) or NTSC (named for the National Television System Committee). For a video encoder to work properly, the termination of the video signal must be set correctly. This is often done via a dip switch to activate the 75-ohm termination.

7.2 Stand-Alone Video Encoders

The most common video encoders are stand-alone versions that offer one or multi-port (often four) connections to analog cameras (Figure 7.4). Multi-port video encoders enable better cost efficiencies, but sometimes the performance and flexibility can be limited. A multi-port encoder is ideal in situations where there are, for example, a few analog cameras located in a remote facility or a place that is a fair distance away from a central monitoring room. Through the video encoder, the video signals

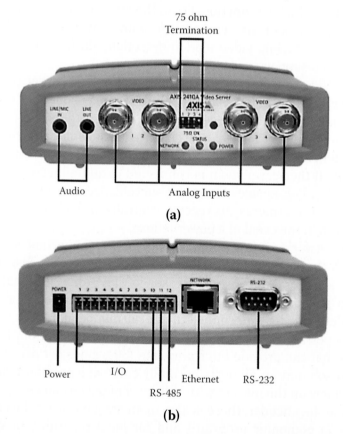

Figure 7.4 (a) Front and (b) back images of a typical stand-alone video encoder with four video inputs, I/O ports, and audio and serial ports.

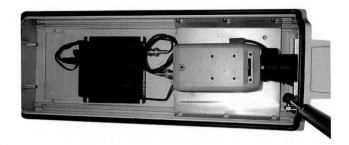

Figure 7.5 A small, single-port video encoder positioned next to an analog camera in a camera housing.

from the remote cameras then can share the same network connection, which dramatically reduces cabling costs.

In situations where investments have been made in analog cameras but coaxial cabling has not yet been installed, it is best to use and position stand-alone video encoders close to the analog cameras (Figure 7.5). This reduces installation costs because it eliminates the need to run new coaxial cables to a central location, as the video can be sent over existing network cabling. It also eliminates the loss in image quality that would occur if video were transferred over long distances through coaxial cabling. With coaxial cables, the quality of video decreases the further the signals have to travel. A video encoder produces digital images, so there is no reduction in image quality due to distance.

7.3 Rack-Mounted Video Encoders

Most companies have a dedicated control room to gather equipment in one location so that operations can be efficiently monitored in a safe and secure environment. In a building containing a large number of analog cameras, this means that vast amounts of coax cabling run to the control room.

If all coax cabling has already been installed and is available from the central room, the installation would benefit from using a video encoder rack with blade video encoders. A blade video encoder is basically a video encoder without a casing. A video encoder in blade version cannot function on its own; it must be mounted in a rack to operate.

A video encoder rack is a chassis that allows a great number of blade video encoders to be mounted in a standard-sized rack, normally 19 inches wide, and managed centrally. Video encoder racks offer functionalities such as an integrated Ethernet switch and hot swapping of blades; that is, blades can be removed or installed without having to power down the

Figure 7.6 An 1U video encoder rack fully populated with three blade video encoders connected with coax cables to twelve analog cameras.

Figure 7.7 Photograph of a blade video encoder (at left) and a 3U video encoder rack (at right). A 3U video encoder rack can hold as many as 48 channels, providing high density and saving on rack space. If a blade needs replacement, having hot-swapping capabilities means that only four analog channels will be down for a short time.

rack. Racks can be of different heights, typically 1U (1 unit = 1.5 inches) or 3U (3 units = 4.5 inches) (see Figures 7.6 and 7.7). A rack typically contains slots for up to 12 interchangeable blade video encoders and provides network, serial communication, and I/O connectors at the rear of each slot, as well as a common power supply. One 3U 19-inch rack typically can support up to 48 channels, providing a high-density solution and saving valuable rack space.

7.4 Video Encoders with PTZ and Dome Cameras

The serial port (RS232/422/485) built into most video encoders is used to control the movement of analog PTZ and dome cameras. Having the video encoder located close to the camera eliminates the need for separate serial wiring from the control board (with joystick and other control buttons) to the PTZ camera, as is the case in an analog CCTV system.

In a network video system, commands from the control board are carried over the same cable as for network video transmission and are forwarded by the video encoder through the serial port to the PTZ camera.

Figure 7.8 An analog dome camera can be controlled via the video encoder using the RS-485 port, making it possible to remotely control it via a joystick.

Video encoders therefore enable control of PTZ functions over long distances using the Internet. A driver must be uploaded to a video encoder to control a specific PTZ camera. Many manufacturers of video encoders provide PTZ drivers for most PTZ and dome cameras. A driver installed on a PC that runs video management software also can be used if the serial port is set up as a serial server that simply passes on the commands.

RS-485 is most commonly used for controlling PTZ functions (Figure 7.8). One of the benefits that RS-485 allows is the possibility to control multiple PTZ cameras using twisted-pair cables in a daisy-chain connection from one dome camera to the next. The maximum distance of an RS-485 cable — without using a repeater — is 1,220 meters (4,000 feet) at baud rates up to 90 kbps (kbps = bits per second).

7.5 Video Decoder

In some installations, there is a need to monitor the network video and audio streams on existing analog monitoring equipment. Using a video decoder, the network video and audio streams are converted back to analog signals, which then can be connected to regular TV sets, analog monitors, and video switches. A typical case could be in a retail environment where the user might want to have traditional monitors in public spaces to demonstrate that video surveillance is used. A video decoder would be used to connect such monitors to a network video stream, coming from either a video encoder or a network camera.

Some video decoders have the ability to decode video from several cameras sequentially. This means that the decoder is decoding video from one camera for five seconds and then automatically changes to the second camera, then the third, etc. This feature allows a guard sitting in front of a monitor to see video from, for example, the five most important cameras (Figure 7.9).

Another common application for video decoders is to use them in an analog-to-digital-to-analog configuration for transporting video over

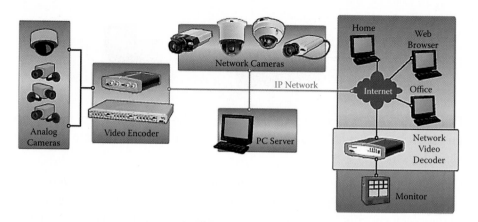

Figure 7.9 With a network video decoder, existing analog monitors can be used to show video and audio from distant cameras that might even be located in a different city.

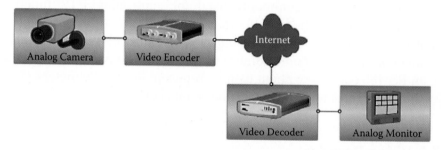

Figure 7.10 An encoder and decoder can be used to transport video over long distances, from an analog camera to an analog monitor.

long distances (Figure 7.10). Distance does not affect the video quality when images are sent in digital format. The only downside may be some level of latency, from 100 milliseconds to a few seconds, depending on the distance and the quality of the network between the endpoints.

7.6 Best Practices

Video encoders offer a valuable solution to the challenge of migrating analog CCTV video to network video. Video encoders play a significant role, particularly in enterprise installations where there may be a large number of analog cameras representing large investments that must be maintained.

It is easy to view video encoders as little more than analog-to-digital converters and, as such, quite straightforward pieces of technology.

However, the demands on video encoders are, in reality, very high, and there are several considerations to make when selecting a video encoder. Considerations to make include:

- *Image quality.* Can the video encoder provide high-quality, de-interlaced digital video? (For more on de-interlacing, see Section 4.4.2.)
- *Resolution.* What resolutions can the video encoder provide?
- *Compression.* What compression standards does the video encoder support?
- *Performance.* How many channels with full frame rate at full resolution can the video encoder provide?
- *Installation.* How easy is the video encoder to install, manage, and upgrade?
- *Rack solution.* Are there rack-mounted versions of the video encoder available?
- *Density.* How many analog channels per rack unit or per Ethernet channel can the system handle?

Video encoders typically fall into the category of products that no one thinks about until something fails. Consequently, reliability and quality are key criteria for video encoders. Video encoders are advanced products that demand careful investigation when making a purchasing decision.

Chapter **8**

Wired Networks

Networks of different types have existed for more than 30 years, providing data exchange between servers and nodes in a computer system. In the early days of office and enterprise networking, many different technologies existed, and only ten years ago there were, for example, Token Ring, Banyan Vines, FDDI (Fiber Distributed Data Interface), and Ethernet. Ethernet has since emerged as the prevailing standardized and established technology and is used to set up networks from home networking to large enterprise systems.

The next three chapters discuss the different aspects of networking. This chapter is designed to give an overview of wired networks, with a focus on Ethernet. Chapter 9 discusses wireless networks, whereas the more technical Chapter 10 focuses on networking technologies, including OSI layers, protocols, and network security. Although the topics are very broad and could easily justify a book each, the content focuses on giving an overview, particularly as it relates to network video.

8.1 The Evolution of Ethernet

Ethernet was originally developed at Xerox PARC in the mid-1970s. In its basic form, the Ethernet standard was released and published by the Institute of Electrical and Electronics Engineers (IEEE) as the IEEE-802.3 standard in 1983. This standard enabled a data transfer rate of 10 Mbps (megabits per second), with individual nodes networked via a coaxial cable in a bus topology. Since then, the basic standard has improved continuously using new transfer media to achieve higher data transfer rates.

Today, Ethernet is based mainly on twisted-pair or fiber-optic cables. The data rates have improved drastically to 100, 1000, and even 10,000 Mbps. In addition, Ethernet is no longer based on a bus topology but, rather, on a star topology, in which the individual nodes are networked with one another via active networking equipment such as switches. A variety of manufacturers offer Ethernet components that can be used to set up very cost-efficient networks. The number of networked nodes on a network can range from two to several thousand. The data rates available depend on the transfer media and networking equipment used in each case. Today's Ethernet networks can certainly provide the performance level required by the most demanding network video applications.

The Ethernet standard is available in many versions. They differ in the transfer medium used and the achievable data rate. The following subsections briefly describe the main versions.

8.1.1 10-Mbps Ethernet

The first Ethernet version was 10BASE5, which used RG-8 coaxial cable. 10BASE2 followed with the 802.3a extension, which used an RG-58 coaxial cable. Such coaxial cable–based networks are not very common today.

The 10BASE-T standard (802.3I) was released in 1990 and uses a twisted-pair cable. A twisted-pair cable is similar to an improved version of a telephone cable. It consists of four pairs of two twisted wires. This is done to improve the electrical properties for data transfer. A 10-Mbps Ethernet uses two of the four pairs of wires to transfer data. The cable is used in combination with RJ-45 plugs and sockets. The maximum length of a cable segment is 100 meters (328 feet), 90 meters (295 feet) of which is the main cable and 10 meters (33 feet) of which is the patch cable (2 × 5 meter patch cable).

Although 10-Mbps Ethernet is still installed and used, such networks often do not provide the necessary bandwidth for some network video applications.

8.1.2 Fast Ethernet

The term "Fast Ethernet" refers to a 100-Mbps Ethernet network and is the most common kind of network today. Fast Ethernet was introduced with the 802.3u extension as a 100BASE-T solution in 1995 and is described in the standard as a variation for twisted-pair cable (100BASE-TX) and glass fiber (100BASE-FX). 100BASE-TX is the most popular Ethernet

Figure 8.1 The standard RJ-45 connector is backwards-compatible and can hence be used in 10 Mbps Ethernet, Fast Ethernet, and Gigabit Ethernet networks.

interface and also provides backwards compatibility with 10BASE-T. Most devices connected on a network, such as a laptop or a network camera, are equipped with a 100BASE-TX/10BASE-T Ethernet interface — most commonly called a 10/100 interface — and hence support both 10-Mbps and Fast Ethernet.

8.1.3 Gigabit Ethernet

The third generation of Ethernet was specified in 1998 by the 802.3z extension, which is based on glass fiber, and in 1999 by the 802.3ab extension for twisted-pair cable (1000BASE-T). Gigabit Ethernet delivers a data rate of 1,000 Mbps (1 Gbps) and is becoming very popular. It is expected to soon replace Fast Ethernet as the *de facto* standard. The main difference, when compared with 10BASE-T and 100BASE-TX, is that 1000BASE-T uses all four pairs of twisted wires in the cable to achieve the high data rates. Most connectors are backwards-compatible with 10-Mbps and Fast Ethernet and are commonly called 10/100/1000 connectors (Figure 8.1).

Various versions such as 1000BASE-SX and 1000BASE-LX are available for use with fiber for transmission over longer distances. SX stands for Short Wavelength and specifies the wavelength to 830 nanometers, whereas LX stands for Long Wavelength, which corresponds to a wavelength of 1,270 nanometers. 1000BASE-SX works with multimode glass fibers (MMF) and permits cable lengths of up to 550 meters (1639 feet). 1000BASE-LX, with multimode glass fibers, can be up to 550 meters long and up to 5,000 m (16,390 feet) long with single-mode fibers (SMF) (see Figure 8.2).

8.1.4 10-Gigabit Ethernet

Gigabit Ethernet is the latest generation and delivers a data rate of 10 Gbps (10,000 Mbps) and is able to use both glass fiber cable as well as

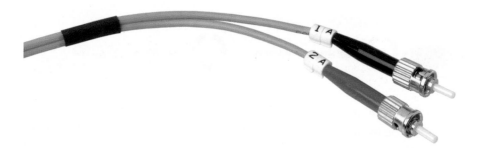

Figure 8.2 Using fiber, longer distances can be bridged. Typically, fiber is used in the backbone of a network and not in nodes such as a network camera.

twisted-pair cable. Glass fiber solutions were specified in 2001 with the 802.3ae extension. 10GBASE-LX4, 10GBASE-ER, and 10GBASE-SR based on glass fiber can be used to bridge distances up to 10,000 meters (6.2 miles). The twisted-pair solution was published in 2006 with the 802.3an specification and permits data transfer of 10 Gbps via twisted-pair cable. All four twisted wire pairs in the cable are used here and a very high-quality cable is required.

10 Gigabit Ethernet is used for backbones in high-end applications that require high data rates (Figure 8.2).

8.2 Network Topologies

Essentially, networks can be built in two different ways: (1) in a star configuration or (2) in a bus configuration. The configuration, also called topology, describes how the individual nodes connect to the network. Although Ethernet can use both topologies, all new networks today use a star topology.

8.2.1 Bus Topology

In a bus topology (Figure 8.3), the transfer medium consists of a main (coaxial) cable connected to all nodes and two terminal resistors. The terminating resistor matches the cable impedance of 50 ohms and prevents reflections. With 10BASE2, the nodes are connected to the network using T-connectors. The transfer medium is available to all nodes at the same time, which means that only one node can send data at a given time, during which the other nodes cannot transmit data.

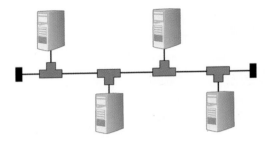

Figure 8.3 In the early days of Ethernet, the bus topology was the most common configuration and may still exist in some older networks.

Figure 8.4 In a star topology, all nodes connect to a central point.

8.2.2 Star Topology

In a star topology (Figure 8.4), the individual nodes are connected in a star formation via a central point, such as a switch. The cables, based on twisted pair or glass fiber, form a point-to-point connection between the connected nodes and the central switch.

In larger network installations, multiple star topology networks are connected in a hierarchy (Figure 8.5). An "uplink" is used to make connections between the network switches. The central parts of the network that connect all the local star topologies often are referred to as the network backbone.

8.3 Network Cabling

In wired networks, all nodes must be connected via some type of network cable. Many different types are available, including coaxial, twisted-pair,

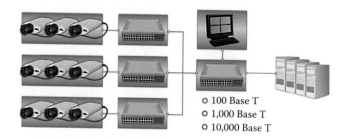

- 100 Base T
- 1,000 Base T
- 10,000 Base T

Figure 8.5 Larger networks are built in a hierarchy, with local star networks connected via a backbone. The higher up in the hierarchy a network is, the higher the network performance.

or glass fiber cable. Each type of cable is available in many different versions. Twisted-pair cables are separated into different categories (Cat). This section outlines the most common types and versions.

8.3.1 Coaxial Cable and BNC

Early versions of Ethernet with 10BASE2 used a coaxial cable called RG-58 (Figure 8.6). RG-58 is similar to the antenna cable on a TV or coax used in analog CCTV (closed-circuit television) installations, but it has a different diameter and resistance. The outer diameter of an RG-58 cable measures 2.95 millimeters and the impedance is 50 ohms, with a BNC (British Naval Connector) at each end. A bus is set up with an RG-58 cable, which is laid from node to node, with the individual nodes are connected to the coaxial cable using a T-connector. A T-connector is basically a T-shaped branch connector, which connects the individual segments of the coaxial bus and connects directly to the node's network interface. Both ends of a coaxial bus then must be terminated with a matching resistor.

The advantage of coaxial cabling is that multiple nodes can be connected together without any central points such as a network switch or hub, and the network can simply be extended. The disadvantages of

Figure 8.6 Ethernet 10BASE2 cabling may still be used in some older installations, but not in new installations today.

Figure 8.7 Twisted-pair cabling includes four pairs of twisted wires, normally connected to a RJ-45 plug at the end.

coaxial cabling are that the data rate is limited to 10 Mbps and faults on the bus can cause the entire network to fail. Due to these drawbacks, coaxial cables are no longer used to set up new network infrastructures.

8.3.2 Twisted-Pair Cable and RJ-45

Twisted-pair cable is, by far, the most common type of cable used in Ethernet networks today (Figure 8.7). The cable consists of eight wires, forming four pairs of twisted wires. Wire pairs, which keep electromagnetic interference low, transfer complementary signals. The maximum length of a twisted-pair cable for Ethernet is normally 100 meters (328 feet).

Depending on the electrical properties of the twisted-pair cable, data transfer of 10, 100, 1,000, or 10,000 Mbps is achievable. In 10- and 100-Mbps Ethernet, only two of the four wire pairs are used for data transfer — that is, one pair of wires for sending data and another pair for receiving data. For 1,000-Mbps (Gigabit) and 10,000-Mbps (10 Gigabit) Ethernet, all four wire pairs are used simultaneously in both directions to reduce the resulting frequencies for the data signals transferred. As cited previously, RJ-45 plugs and sockets are used as connectors.

8.3.3 Cable Categories

By supporting different transfer frequencies (measured in megahertz [MHz]), different data rates (measured in Mbps) are supported. Different types of cables, referred to as categories or Cat, are required to support certain transfer frequencies. The individual categories are defined in the ISO/IEC-11801 standard and specify certain transfer properties for the twisted-pair cable, such as impedance, bandwidth, damping, and near-end crosstalk (NEXT). The categories relevant for Ethernet are Cat-3 to Cat-7, as explained below.

- *Cat-3* is a twisted-pair cable for transfer frequencies up to 10 MHz, which corresponds to the requirement of 10-Mbps Ethernet (10BASE-T). Due to the limitation of 10 MHz or 10 Mbps, Cat-3 should no longer be installed today.
- *Cat-5* is a twisted-pair cable for transfer frequencies up to 100 MHz which corresponds to the requirement of Fast Ethernet (100BASE-TX). Cat-5 exists in many installations today and is still used in some new installations.
- *Cat-5e* is a twisted-pair cable for transfer frequencies up to 100 MHz and is built to the same specs as a standard Cat-5 cable. By extended inspection measurements and to make sure that the cable complies with the requirements for operating Gigabit Ethernet (1000BASE-T), the "e" is added to the categorization of the cable.
- *Cat-6* was originally specified for a twisted-pair cable with transfer frequencies up to 250 MHz. With the introduction of 10GBASE-T, this category was extended and divided into subcategories. *Cat-6a*, with transfer frequencies of up to 625 MHz, and *Cat-6e*, of up to 500 MHz, were added, which met the requirements of 10GBASE-T. Cat-6a permits distances of 100 meters and Cat 6e permits only 55 meters.
- *Cat-7* is a twisted-pair cable with transfer frequencies of up to 600 MHz, which corresponds to the requirements of 10GBASE-T.

8.3.4 Twisted-Pair Cable Types

The primary difference among the different categories explained above lies in whether or not the cable is equipped with shielding and, if so, the type of shielding used. The shield works as an electromagnetic protection for the twisted pairs within the cable, which not only improves the performance of the cable but also adds to the manufacturing cost. The following are the main types:

- *UTP cable* (unshielded twisted-pair cable) is a cable that uses no shield whatsoever. UTP cables are typically covered only with a plastic cover for physical protection of the cables.
- *STP cable* (shielded twisted-pair cable) is a cable with one metallic shield covering all twisted-pair cables.

- *S-UTP cable* (screened unshielded twisted-pair cable) is a cable similar to the unshielded twisted-pair cable but wrapped in a shielding foil with an integrated wire, which helps to shield the cable.
- *U/FTP cable* (unshielded/foiled twisted-pair cable) is a cable in which each of the pairs of wires is shielded, but there is no overall shielding.
- *S/FTP cable* (screened/foiled twisted-pair cable) is a cable in which each of the pairs of wires is shielded with metal foil and an additional overall shield made of metallic foil and copper wire is wrapped around all the pairs of wires.

The higher the transfer frequency, the better the shielding required to ensure that the twisted-pair cable does not transmit any electromagnetic interference and that the data transfer is error free.

8.3.5 Glass Fiber Types

A glass fiber cable is quite common in today's high-performance networks. Although somewhat more expensive, it offers several advantages. The most obvious is increased segment lengths of more than 100 meters (328 feet). Additionally, fiber is not sensitive to electromagnetic interference. The latter can be an advantage in industrial environments. In a network video application, fiber can make it possible to place a network camera outside a building, for example, in a parking lot located farther away from the building. Fiber cables are normally used in the backbone of a network due to their high performance.

When using glass fibers, light is utilized as an information carrier. It uses standard LEDs (light emitting diodes) or special laser LEDs to transmit and receive light at the two ends of the fiber. The LEDs use wavelengths of 850, 1,300, or 1,550 nanometers; and depending on the glass fiber type, the light spreads through the glass fibers in different modes.

The two modes are multimode fiber (MMF) and simple-mode fiber (SMF). For MMF, the core diameter is 50 micrometers or 62.5 micrometers, and the outer diameter is 125 micrometers. For SMF, the core diameter is between 9 and 10 micrometers, and the outer diameter is 125 micrometers. Generally speaking, shorter distances can be bridged with MMF and longer distances with SMF. The bridgeable distance depends on the damping and dispersion. In the case of MMF, the bandwidth also affects the distance that can be bridged. Similar to twisted-pair cables, glass fiber cables defined by EN 50173 and ISO/IEC-11801 are divided

Table 8.1 Glass Fiber: Categories and Performance Characteristics

Fiber Type	OM1	OM2	OM3	OS1
	50/125 μm 62.5/125 μm	50/125 μm 62.5/125 μm	50/125 μm	9 to 10/ 125 μm
Damping				
850 nm	3.5 dB/km	3.5 dB/km	3.5 dB/km	Not def.
1300 nm	1.5 dB/km	1.5 dB/km	1.5 dB/km	1 dB/km
1550 nm	Not def.	Not def.	Not def.	1 dB/km
Bandwidth length product				
850 nm	200 MHz x km	500 MHz x km	1500 MHz x km	Not def.
1300 nm	500 MHz x km	500 MHz x km	500 MHz x km	Not def.

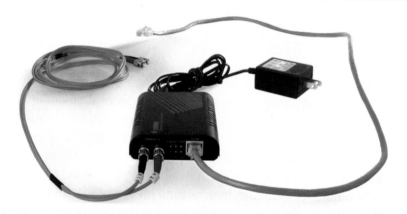

Figure 8.8 A media converter, also called a fiber transceiver, can be used to connect any device with an RJ-45 plug (e.g., a network camera) to a fiber network.

into different categories. Table 8.1 shows the various categories and performance characteristics of glass fiber.

When used in an Ethernet environment, Duplex SC or Duplex ST is used as the connectors for glass fiber. Most network video products, such as network cameras, only come with a twisted-pair interface. To connect them to a fiber cable, a media converter must be used (Figure 8.8).

8.4 The Basics of Ethernet

The basic idea behind an Ethernet network is that the network provides a medium like the air (also called ether, and therefore the name Ether-net) through which all nodes can communicate. Although many things have changed throughout the years as Ethernet has substantially improved, some of the basic elements still remain.

8.4.1 MAC Addresses

Media Access Control (MAC) addresses are used as source and destination addresses. A MAC address has a 48-bit address space, which allows for potentially 2^{48} or 281,474,976,710,656 possible MAC addresses. These are unique addresses in hexadecimal format with a length of six bytes (e.g., 00-40-8C-18-32-78), where the first three bytes describe the manufacturer of the equipment and the last three correspond to the serial number of the specific networking device (Figure 8.9). The manufacturer sets the MAC address and the user cannot change it. Each network device has a unique MAC address, which is often printed on the backside of the device. MAC addresses are always used when sending data from one device to another on a network, whether in a LAN (local area network) or over the Internet. For communication within a local network, MAC addressing may be all that is needed. In many cases, however, IP addresses are required in addition to MAC addresses. For more information on sending data packets, see Chapter 10.

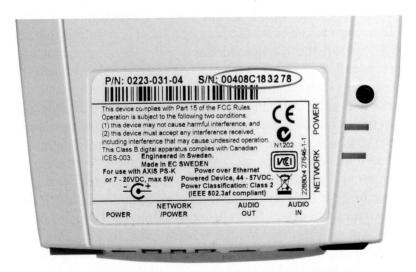

Figure 8.9 Each networking device has a unique MAC address, which is often found on the product label. In this photo, it is shown as the serial number (S/N).

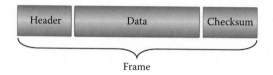

Frame

Figure 8.10 A typical frame consists of the header, the data, and the check-sum. This means that only part of the bits transferred over a network is actual data. The remainder is called "overhead."

8.4.2 Frames

For transferring data, the data is packaged into what is known as a frame. A frame consists of a header, the data that will be transferred, and a checksum, via which the recipient can recognize transmission errors (Figure 8.10). The key information in the header consists of the destination and source addresses of the frame. The destination address specifies the node the frame should be sent to, and the source address specifies the node from where the frame is sent.

8.4.3 Carrier Sense Multiple Access/ Collision Detection (CSMA/CD)

An Ethernet network uses a shared medium as a communication channel; thus, all the nodes connected to it can initiate a data transfer at any given time. To accommodate this, an access protocol is required that enables controlled use of the transfer medium. The access protocol used for Ethernet is called CSMA/CD (Carrier Sense Multiple Access/Collision Detection) and works as follows.

Prior to sending data, all nodes check whether the transfer medium is free, that is, that no other communication is occurring. If the medium is free, the node starts to send data. During transmission, the node continues to listen to determine whether the data can be transferred without collision.

If another node starts to send data at the same time, the frames will collide on the network, making the transmission unsuccessful. If this happens, the nodes will detect the collision and stop the transmission. Then they will wait before restarting the transmission. The waiting period at each node is generated randomly so that the nodes avoid accessing the medium at the same time to prevent a collision from occurring again.

The area of a network where collisions can occur is known as a collision domain. The probability of collisions depends on the number of transmitting nodes. Many networks have many, fairly passive nodes (e.g., display terminals that mainly receive video). These nodes do not contribute to the probability of collisions in a significant way.

8.4.4 Half and Full Duplex

Half duplex means that data can be sent only in one direction at a time, whereas full duplex means that data can be sent and received simultaneously, thus increasing the performance.

Coaxial cables do not provide separate transmission and receiving channels and, thus, only support half duplex. Twisted-pair cables and glass fiber cables have separate transmission and receiving channels and, therefore, support full duplex. If an interface operates in full-duplex mode, data can be transferred via the medium without using CSMA/CD.

Ethernet products with twisted-pair interfaces today normally support several data rates in half- or full-duplex mode. A twisted-pair interface can adjust itself automatically to the data rate and transfer mode using what is known as auto-negotiation. Two nodes that are connected will negotiate automatically to find the highest common data rate and transfer mode and then automatically set this at both nodes. If for some reason the negotiation fails, many devices still have the possibility to set the transfer mode manually (Figure 8.11).

8.5 Networking Equipment

To set up a direct network connection between two nodes via a twisted-pair cable, a so-called crossover cable can be used. The crossover cable simply crosses the transmission pair on one end of the cable with the receiving pair on the other end, and vice versa.

To network multiple nodes, network equipment — the most common being the network switch or, in older networks, a network hub — is required. When using a network switch, a regular network cable is used instead of a crossover cable. There is also other commonly used network equipment, such as routers, for accessing the Internet.

8.5.1 Hubs

A hub, also called a repeater, is the simplest type of networking equipment (Figure 8.12). It works on the first layer of the OSI model (physical layer; see Chapter 10). All nodes are connected to the hub, forming a

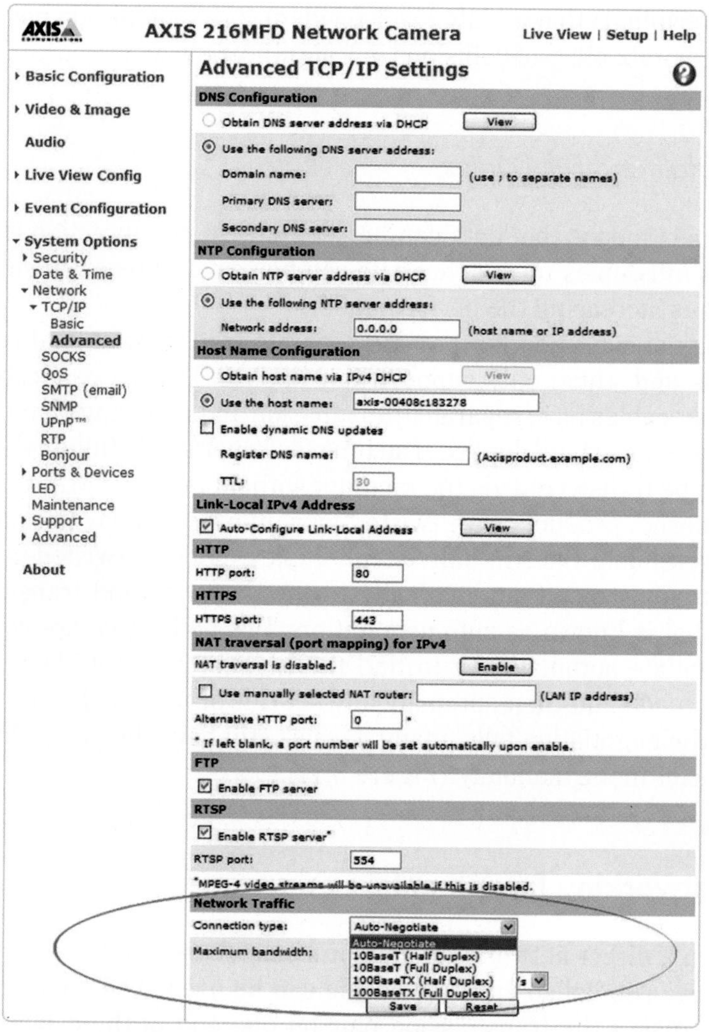

Figure 8.11 In most cases, the network device should be set to auto mode, which will automatically detect if full or half duplex can be used. In some cases, it is beneficial to manually set the mode to half or full duplex.

collision domain. Inside this collision domain, only one node can send data while all other nodes receive data at the same time. If another hub is connected, the collision domain is extended. If a hub receives data on one connection, it sends out the data to all the other connections.

In a hub environment, all nodes operate in half-duplex mode. Additionally, a classical hub can only support one data rate at a time, either at 10 or 100 Mbps, meaning that all nodes in the network must support the same data rate. One exception here is the dual-speed hub,

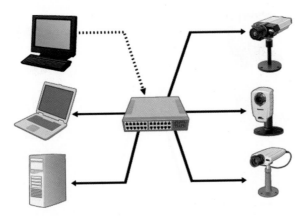

Figure 8.12 A hub is the simplest form of networking equipment. It makes it possible for several nodes to communicate with all other nodes that are connected to the hub.

which supports two data rates. Hubs are rarely used in modern networking. Switches are used instead.

8.5.2 Switches

The network switch includes more intelligence and can forward the network traffic in a much more efficient manner than hubs. The switch drastically improves the performance of a network and also the security because data is not always sent to all nodes on the network.

Whereas the network hub operates on Layer 1 in the OSI model, the network switch manages the data also on the second layer and, in some cases, the third layer in the OSI model. That is why some switches today are referred to as "Layer 3 switches." For more information on the OSI model, see Chapter 10.

Different vendors have used the term "Layer 3 switch" to describe different functionality, which has confused the market. Being a Layer 3 switch means that the switch includes the function of a router (see next subsection). There are also other important Layer 3 functionalities that are especially relevant to managing video. One is Quality of Service (QoS), which is required in managing the bandwidth. Another is IGMP (Internet Group Management Protocol) snooping, which is useful in multicasting networks. Chapter 10 provides more information about these technologies.

Data forwarding in a switch takes place via a learning process. The switch registers the address of the connected device. When the switch receives data, it forwards it only to the port connected to the device with

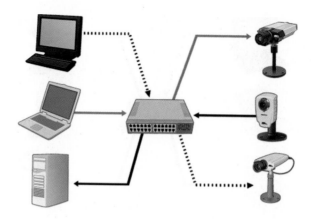

Figure 8.13 In a network switch, data transfer is managed very efficiently because data traffic can be directed from one device to another without affecting any other ports on the switch.

the appropriate destination address (Figure 8.13). For more on port forwarding, see Chapter 10.

Switches typically indicate their performance in per-port rates and in backplane or internal rates (both in bit rates and in packets per second). The port rates indicate the maximum rates on specific ports. This means that the speed of a switch (e.g., 100 Mbps) is often the performance of each port.

A network switch normally supports different data rates simultaneously. The most common rates used to be 10/100, supporting 10 Mbps as well as Fast Ethernet. However, 10/100/1,000 are quickly taking over as the standard switch, thus supporting 10-Mbps Ethernet, Fast Ethernet, and Gigabit Ethernet simultaneously. The transfer rate and mode between a port on a switch and a connected device are normally determined through auto-negotiation, whereby the highest common data rate and best transfer mode are used. A switch also allows a connected device to function in full-duplex mode, that is, to send and receive data at the same time, resulting in increased performance.

8.5.3 Routers

A network router is a device that routes information from one network to another. A router is used primarily for connecting a local network to the Internet. It forwards only the data packages that are supposed to be transmitted over to another network.

A router can forward data between completely different network technologies, thus creating a larger interconnected network (internets).

Traditionally, routers were — and sometimes still are — referred to as gateways.

8.5.4 Firewalls

Firewalls are designed to prevent unauthorized access to or from a private network. Firewalls can be implemented in both hardware and software, or a combination of both. Firewalls frequently are used to prevent unauthorized Internet users from accessing private networks connected to the Internet. Messages entering or leaving the Internet pass through the firewall, which examines each message and blocks those that do not meet the specified security criteria.

8.5.5 Internet Connections

To connect a network to the Internet, a network connection via an Internet service provider (ISP) must be established. With the necessity of Internet in many operations, Internet connection speeds are increasing and costs are decreasing.

When connecting to the Internet, terms such as "upstream" and "downstream" are used. "Upstream" describes the transfer rate with which data can be uploaded from the device to the Internet, such as when video is sent from a network camera. "Downstream" is the transfer speed for downloading files, such as when a monitoring PC receives video.

In most applications (e.g., a laptop connecting to the Internet), downloading information from the Internet is the most important speed to consider. In a network video application with a network camera at a remote site, the upstream speed is more relevant because data (video) from the network camera will be uploaded to the Internet.

Some of the more common technologies for Internet connections include:

- *DSL.* Digital Subscriber Line is the technology used to overlay data traffic on a regular telephone connection that uses copper wires. Data speeds of 256 Kbps to 1 Mbps are the most common.
- *ADSL.* Asymmetric DSL is a DSL connection where the bandwidth is asymmetric, that is, higher in one direction than the other. Most often, the downstream speed is considerably higher (five to ten times) than the upstream speed. ADSL connections, with their higher downstream speeds, are perfect for a PC that is

surfing (downloading information from) the Internet. If a network camera is connected to an ADSL connection, the upstream speed should be higher, which unfortunately is not very common.

- *SDSL.* Symmetric DSL is a DSL connection with the same upstream and downstream bandwidth. This type of connection is therefore more suitable for a network camera than an ADSL connection.
- *Cable.* A cable connection uses the cable TV connection. It means overlaying the data communication on the cable used for TV.
- *T1.* A T1 connection provides a bandwidth of 1.544 Mbps for both upstream and downstream data transfer. Several T1 connections can be pooled via a device that provides 3-, 6-, and 9-Mbps connections.
- *Fiber.* A fiber connection means that a glass fiber connects directly into an ISP's network, providing data transfer speeds of 10, 40, or even 100 Mbps to the Internet.

8.6 Power over Ethernet

Power over Ethernet (PoE) provides the option of supplying devices connected to an Ethernet network with power using the same cable as for data communication. The whole idea stems from older telephone systems where the telephone line was both a means for communication and the power source for the telephone. In the early days of IP telephony, the same functionality was mimicked using Ethernet cabling. Today, PoE is widespread; it is used to power IP phones, wireless devices, and network cameras in Ethernet networks.

The primary benefit in using PoE is the inherent cost savings. The fact that no power cable must be installed can save — depending on the camera location — up to a few hundred dollars per camera. It also makes it easier to move a camera to a new location or add cameras to a video surveillance system.

Additionally, PoE can make a video system more secure. A video surveillance system with PoE can be powered from the server room, which is often backed up with a UPS (uninterruptible power supply). This means that the video surveillance system can be operational even during a power outage.

8.6.1 The 802.3af Standard

Most PoE devices today conform to the 802.3af standard, which was published by the IEEE in 2003. The 802.3af standard uses standard

Cat-5 or better cables and ensures that data transfer is not affected and that the maximum possible length of a twisted-pair cable segment is 100 meters (328 feet). In this standard, the device that supplies the power is referred to as the "power sourcing equipment" (PSE), whose functionality can be built into a network switch or provided by a midspan (see Section 8.6.2). The device that receives the power is referred to as a "powered device" (PD). The functionality is normally built into a network device such as a network camera or provided in a stand-alone splitter (see Section 8.6.2).

Backwards-compatibility to non-PoE-compatible network devices is guaranteed. The standard includes a method for automatically identifying if a device supports PoE, and only when that is confirmed will the power be supplied to the device. This also means that the Ethernet cable connected to a PoE switch will not supply any power if it is not connected to a PoE device. This eliminates the risk of getting an electrical shock when installing or rewiring a network.

In a twisted-pair cable, there are four pairs of twisted cables. PoE can use either the two dormant wire pairs or overlay the current on the wire pairs used for data transmission. As a rule, switches with built-in PoE supply electricity through the two pairs of wire used for transferring data, whereas midspans normally use the two dormant pairs. A PD supports both options.

According to 802.3af, a PSE provides a voltage of 48 volts DC at a maximum power of 15.4 watts per port. Considering that power loss takes place on a twisted-pair cable, only 12.95 watts is guaranteed for a PD. The 802.3af standard specifies various performance categories for PDs. Table 8.2 provides the various classes and power ranges according to the 802.3af standard.

PSEs such as switches and midspans only supply a certain amount of power, typically 300 to 500 watts. On a 48-port switch, that would mean

Table 8.2 Classes and Power Ranges of Powered Devices According to the 802.3af Standard

Class	Power Range (Watt)
Class 0	0.44–12.95
Class 1	0.44–3.84
Class 2	3.84–6.49
Class 3	6.49–12.95
Class 4	For future use

6 to 10 watts per port if all ports are connected to devices that use PoE. Unless the PDs support power classification, the full 15.4 watts must be reserved for each port that uses PoE, which means a switch with 300 watts can only supply power on 20 of the 48 ports. However, if all devices let the switch know that they are Class 1 devices, the 300 watts will be enough to supply power to all 48 ports.

Most fixed network cameras can receive power via PoE and are normally identified as Class 1 or Class 2 devices. PTZ (pan, tilt, zoom) and dome cameras with motor control, and cameras with heaters and fans, require more power than can be supplied by the current 802.3af PoE standard.

8.6.2 Midspans and Splitters

With PoE, two new network devices have been introduced, namely the midspan and the splitter, the latter also called an active splitter (Figure 8.14). Both devices are used to enable an existing network to support PoE.

The midspan, which adds power to an Ethernet cable, is placed between a network switch and powered devices. Midspans with 1, 6, 12, 24, or 48 ports that support the 802.3af standard are commonly available.

A splitter, or active splitter, is used to split the power and data in an Ethernet cable into two separate cables, which then can be connected to a device that has no built-in support for PoE. Because the 802.3af standard supplies only 48 volts DC, another function of the splitter is to step down the voltage to the appropriate level for the device, often 12 or 5 volts.

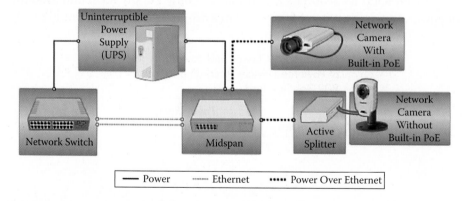

Figure 8.14 Using midspans and splitters, existing systems can be upgraded with PoE functionality.

8.6.3 Future Standards

The 802.3af standard provides a maximum guaranteed power of 12.95 watts at the PD. Because of all the obvious benefits of PoE, there has been ongoing work to extend the amount of power that can be distributed over a network so that devices that require more power — such as laptops, network dome cameras, and cameras that require the use of fans and heaters — also can benefit from PoE. There are already some proprietary products on the market that provide more power.

The IEEE, as a consequence, is working on a new version of PoE, often referred to as HiPoE, PoE+, or PoE Plus, which is to be specified by the 802.3at extension of the standard. The 802.3at standard should offer a solution that delivers 30 watts of power via two pairs of wires and 60 watts via four pairs of wires. The final specifications are still to be determined, and the standard is expected to be ratified in late 2008.

8.7 Virtual Local Area Networks (VLANs)

When designing a network video system, there is often a desire to keep the network separate from other networks, for both security and performance reasons. At first glance, the obvious choice would be to build a separate network. Although the design would be simplified, the cost of purchasing, installing, and maintaining the network would often be higher than that using a technology called virtual local area network (VLAN).

VLAN is a technology for virtually segmenting networks, a functionality that is supported by most network switches today. It can be achieved by dividing network users into logical groups. Only users in a specific group are capable of exchanging data or accessing certain resources on the network. If a network video system is segmented into a VLAN, only the servers located on that VLAN can access the network cameras (Figure 8.15).

The primary protocol used when configuring VLANs is IEEE 802.1Q, which tags each frame or packet with extra bytes to indicate the virtual network to which the packet belongs. Before this standard became available, several manufacturers had proprietary VLAN protocols. VLANs operate on Layer 2 of the OSI model. (More information on OSI is available in Chapter 10.)

The four ways to assign VLAN memberships are:

1. *Port based*, which means that a physical port on a switch is configured as part of a VLAN or not

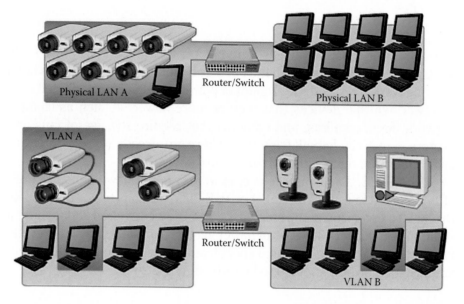

Figure 8.15 Top: Two networks physically separated. Bottom: The router or switch manages the IP addresses, bandwidth, and security allocated to those on VLAN A (with access to video) and VLAN B (general-purpose traffic). No matter where users are physically located, all those on VLAN A will have access to video, but those on VLAN B will not.

2. *MAC based*, which means that each MAC address that is part of a VLAN is listed in the switch

3. *Protocol based,* which means that Layer 3 data is used to determine which VLAN a frame belongs to; for example, AppleTalk is one VLAN and IP, another

4. *Authentication based*, which means that devices are placed in a VLAN based on 802.1X authentication (See Chapter 10 for more information on 802.1X.)

8.8 Best Practices

Whereas network bandwidth was a scarce resource in the early days of network video, bandwidth in today's networks is plentiful. Nonetheless, there are several important considerations to make when designing a network appropriate for a network video application.

- *Is existing cabling available?* If so, this can save a lot of money. Make sure it is of appropriate quality to support the selected network speed and PoE, if used.

- *What category cable should be used?* If there is no existing cabling, new cable should be installed. When looking at the cost of installing the cable, the actual cable is normally a smaller portion of the total cost, with labor being a bigger part. The recommendation is to choose as good a cable as possible to future-proof the installation. At least Cat-5e or Cat-6 cables should be used. For backbones, Cat-6 or Cat-7 cables should be used.
- *What cable distances are needed?* If the distance is greater than 100 meters (328 feet), a fiber cable may be the best solution.
- *Is it possible to use PoE?* PoE provides huge savings, so make sure it is used for as many devices as possible. If PoE is used, make sure the power available in the switch or midspan is sufficient for the nodes connected and that the nodes support power classification.
- *What bandwidth is required?* Most switches today are 10/100/1,000, which provide more bandwidth than is normally required. To future-proof a network, it is a good idea to design a network such that only 30 percent of its capacity is used.
- *Should there be a WAN connection in the system?* The WAN (wide area network) bandwidth is normally limited. The system should be designed so that it does not overload the WAN bandwidth.
- *Should there be a VLAN or separate network?* VLANs normally provide a better and more cost-efficient solution than a separate network.

A rule of thumb always is to build a network with greater capacity than is currently required. Because more and more applications are running over networks today, higher and higher network performance becomes a requirement. Whereas network switches are easy to upgrade after a few years, cabling is normally much more difficult to replace.

Chapter 9

Wireless Networks

Wireless networks have quickly become very popular, especially in home and small office networking. For corporate networking, the adoption rate has been somewhat slower, primarily due to concerns about network security. For video surveillance applications, wireless technology presents an interesting option to wired networks, mainly due to the flexibility and cost savings. For example, installing a wireless camera in a parking lot could be cheaper and easier than the alternative of pulling a cable through the ground.

In the early days of wireless networks, security and data rates or available bandwidth were concerns that were especially relevant to IP (Internet Protocol)-based video applications. However, rapid development within the IT industry has made the wireless networking infrastructure a very interesting alternative in network video applications. It offers a way to cost-efficiently and quickly deploy cameras over a large area, especially in city center surveillance applications.

Wireless networks also enable mobility; that is, a device such as a network camera can move freely inside a wireless network's coverage range. This benefit is particularly useful for camera installations in vehicles. It means that a camera in a bus, for example, can be accessed live from a remote location. It is also relatively simple to extend the network because anyone within range of the network and with the appropriate log-in and authorization data can access the network wirelessly. In older buildings under the protection of the Heritage Act, wireless networks may be the only alternative because standard network cables cannot be installed.

All network cameras can be integrated into a wireless network through the use of a camera with built-in wireless capabilities (Figure 9.1) or by using a wireless bridge (Figure 9.2).

Figure 9.1 A network camera with built-in wireless networking — in this case, 802.11b/g.

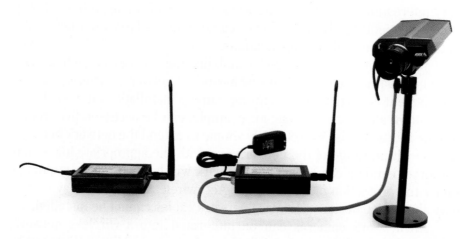

Figure 9.2 Using a wireless bridge, any network camera can interface with and benefit from a wireless network.

Wireless networks come in many different forms. Wireless local area networks or WLAN technology are predominantly based on the IEEE 802.11 standard. There also are other standards as well as proprietary technologies available, some of which are interesting for video surveillance applications. Wireless networks can be designed in different ways, ranging from point-to-point, point-to-multipoint, and mesh networks. Different wireless networks operate in different frequency ranges, also called spectrum. One of the benefits of 802.11 wireless standards is that they all operate in a license-free spectrum, which means that there is no license fee associated with setting up and operating the network.

9.1 The Basics of Wireless Networks

Wireless networks involve the transmission of electromagnetic waves through the air. Ever since the Italian inventor Marconi conducted the first wireless transmission of Morse code in 1895, wireless technology has played an increasingly important role in people's lives. Although the basic technology remains the same, many developments have been made during the past hundred years, resulting in the widespread use of wireless cell phones and the ability to listen to music on the radio from almost anywhere.

9.1.1 Wireless Spectrums

Wireless communication is conducted at certain frequencies measured in hertz (Hz), which is the unit for (electromagnetic) cycles per second. Using different frequencies, different wireless systems have the ability to communicate at the same time. This is why a person can communicate, for example, on his or her cell phone and at the same time listen to the radio — because those two systems are using different frequencies. If they were using the same frequency, there would be interference between the systems.

That is why regulatory bodies manage radio frequencies, also called spectrum. Some frequencies are in the licensed spectrum, which means that governmental bodies such as the FCC (Federal Communications Commission) must approve the use of certain frequency bands. Other parts of the spectrum are license free, which means that anyone can use them without prior approval and at no cost.

The lower the frequency, the longer reach the radio signal has. On the flip side, higher frequencies enable greater amounts of data throughput (bandwidth). How the different frequency ranges are divided and allocated for different uses differs in different regions of the world. Table 9.1 provides the spectrum allocation in the United States. Most of the wireless spectrum between 300 MHz and 3 GHz is allocated today. Below 300 MHz, applications such as FM radio (88–110 MHz) occupy the spectrum.

9.1.2 Signal Strength

There are different ways to measure signal strength in a wireless system. One way is to examine the amount of radio-frequency (RF) energy transmitted. The more energy, the further the signal will reach. The unit for energy is watts (W), and in most wireless systems, the energy is in the milliwatt (mW) range. For example, in a typical 802.11 system, the maximum energy is 100 milliwatts at the wireless access point. At higher energy levels, the radio signal will reach further but the risk of interference is also greater.

However, it turns out that measuring RF energy in milliwatt units is not always convenient. This is due to the fact that a signal's strength does not fade in a linear manner, but inversely as the square of the distance. This means that if the signal level is 100 at a certain distance from an access point, then the signal level at a distance twice as far away will decrease by a factor of four, that is, to 25. The fact that exponential measurements are involved in signal strength measurement is one reason why the use of a logarithmic scale of measurement was developed as an alternative way of representing RF power. The dBm (dB milliwatt) is a logarithmic measurement of signal strength, and dBm values can be exactly and directly converted to and from milliwatt values as per the formula below:

$$dBm = \log(mW) \times 10$$

Therefore, 100 mW = 20 dBm.

If the signal strength in milliwatts (mW) is cut in half, the signal strength is reduced by 3 dBm. There are also other ways to measure signal strength. The IEEE 802.11 standard defines a mechanism by which RF energy can be measured using the circuitry on a wireless NIC (network interface card). This numeric value is an integer with an allowable range between 0 and 255 (a 1-byte value) called the Receive Signal Strength Indicator (RSSI). Different vendors interpret the RSSI value in different ways.

9.1.3 Antennas

The antenna is an important part of a wireless system. The better the antenna, the longer the range of the wireless connection. The

Table 9.1 Wireless Spectrum Allocation in the United States

	Spectrum Allocation	Application
300 MHz	322–450	Fixed, Mobile, Satellite, Navigation, Weather Aids
	450–470	Private Land Mobile
	470–698	Core UHF TV Bands Channels 14–51 (Excluding Ch. 37)
	608–614	Radio Astronomy
	698–700	Reallocated TV Channels 52–59 (Lower 700 MHz)
	746–806	Reallocated TV Channels 60–69 (Upper 700 MHz)
	824–849	Cellular
	869–894	Cellular
	902–928	Unlicensed Band
	929–960	Fixed, Mobile
1000 MHz	960–1215	Navigation
	1215–1240	Radar
	1225.5	GPS
	1300–1350	Navigation
	1350–1390	Fixed, Mobile, Radar, Navigation (GPS 1381.05 MHz)
	1400–1427	Weather Aids
	1435–1525	Aeronautical Telemetry
	1525–1559	L-Band MSS Downlink (Possible MSS Flex)
	1559–1610	Satellite, Weather Aids, Radio Astronomy, Navigation
	1575.42	GPS
	1610–1700	Weather Satellites
	1710–1755	Transfer Band (Possible 3G)
	1755–1850	Government Fixed and Mobile (Possible 3G)
	1850–1910	PCS
	1910–1930	Unlicensed PCS (Possible 3G)
	1930–1990	PCS
2000 MHz	1990–2025	2 GHz MSS Uplink (Possible MSS Flex, Possible 3G)
	2025–2110	Space, Fixed, Mobile (Broadcast Auxiliary)
	2110–2150	Fixed and Mobile (Possible 3G)
	2165–2200	2 GHz MSS Downlink (Possible MSS Flex, Possible 3G)
	2200–2290	Fixed, Mobile, Satellite
	2290–2320	WCS
	2320–2345	SDARS
	2345–2360	WCS
	2400–2500	Unlicensed Band
	2500–2655	MDS - ITFS (Possible 3G)
3000 MHz	2700–3100	Navigation

performance, or gain, of an antenna is measured in dBi (decibel isotropic), which describes the gain a given antenna has over a theoretical isotropic (point source) antenna, which normally ranges from 1 to 20 dBi. The ERP (effective radiated power) is defined as the effective power of the transmitter antenna. It is equal to the sum of the antenna gain (in dBi) plus the power (in dBm) into the antenna. For example, a 12-dBi gain antenna fed directly with 15 dBm of power has an ERP of:

$$15 \text{ dBm} + 12 \text{ dBi} = 27 \text{ dBm} \ (500 \text{ mW})$$

A longer antenna normally has a higher gain. The appropriate length of an antenna depends on the frequency because the antenna should be a multiple of the wavelength to enable the highest efficiency.

The two main types of antennas are omnidirectional antennas and directional antennas (Figure 9.3). An omnidirectional antenna spreads the electromagnetic waves in all directions, which makes it appropriate for multipoint networks where wide coverage is desired. A directional antenna focuses the electromagnetic waves in one direction, enabling the radio signal to reach a longer distance but only in one direction. Directional antennas are appropriate in point-to-point applications and come in different types, such as Yagi antennas, patch antennas, and parabolic grid antennas; many have gains between 10 and 30 dBi.

9.1.4 Radio Wave Propagation

Although open air is best for transmitting radio signals, radio signals do not require free line-of-sight for transmission. This means that all radio systems are so-called "non–line-of-sight" systems, as opposed to optical transmission systems where the receiving and the transmitting end must be able to "see" each other. Some systems are better at reaching further and through certain materials, as explained below.

The wireless signal propagates best through open air but can technically go through most media such as wood and concrete — that is, through buildings. Buildings consisting of a lot of metal provide some difficulties for radio signals because the radio signals reflect off metals rather than penetrate through them. If a building consists of a metal with high conductivity, such as a copper material without holes, virtually no wireless signal can penetrate it. Such a building or room is called a "Faraday cage" after the electromagnetic researcher, Michael Faraday.

Radio signals of lower frequencies and longer wavelengths propagate further than signals of higher wavelengths. Lower radio frequencies

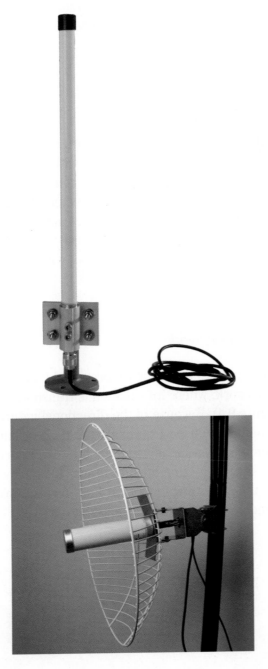

Figure 9.3 An example of an omnidirectional (top) and a directional (bottom) antenna.

also have a better ability to penetrate through materials such as wood or concrete. This is why a signal from an AM radio station can be received on the other side of an ocean, whereas FM radio channels need to be switched every few miles when on the road.

9.2 Wireless Network Architectures

Depending on the application, wireless networks can be designed in different ways. In some scenarios — for example, with a single camera located in a parking lot — a point-to-point wireless connection makes the most sense for sending video back to the building where the video surveillance system is located. In a city center surveillance application with hundreds of cameras, a point-to-multipoint or even a mesh wireless network may be the appropriate solution. These three types of wireless architectures are explained in more detail below.

9.2.1 Point-to-Point Network

A point-to-point network (sometimes abbreviated PTP or P2P) is the simplest wireless network, where information is transmitted from one point to another (Figure 9.4). Because of the directional nature of the system, directional antennas are used to provide the highest bandwidth for the link. The system also can be adjusted for minimal interference and the highest level of security. If line-of-sight is available, it also would be possible to use a high-performance optical link, for example, between two buildings located on opposite sides of a highway.

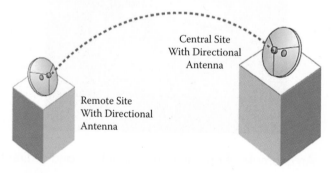

Central Site
With Directional
Antenna

Remote Site
With Directional
Antenna

Figure 9.4 A point-to-point wireless network.

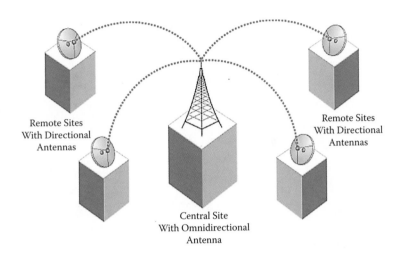

Remote Sites
With Directional
Antennas

Remote Sites
With Directional
Antennas

Central Site
With Omnidirectional
Antenna

Figure 9.5 A point-to-multipoint wireless network.

9.2.2 Point-to-Multipoint Network

A point-to-multipoint network (sometimes abbreviated PTMP or P2MP) is the most common type of wireless network (Figure 9.5). An example of such a network is an FM radio station transmitting radio signals to many radio receivers. Because of the nature of the network, the central point uses an omnidirectional antenna. The surrounding points use a directional antenna unless they are mobile (e.g., a car), in which case an omnidirectional antenna would be preferred. A point-to-multipoint network can be of broadcast and simplex nature, as in the FM radio case, or provide full duplex with data being sent in both directions, as in a regular wireless LAN (local area network) application.

9.2.3 Mesh Network

A wireless mesh network (Figure 9.6) is characterized by several connection nodes that provide individual and redundant connection paths between one another. To accommodate this, special routing protocols that guarantee data exchange via the most appropriate connection path are used. When selecting a path, factors such as bandwidth, transfer errors, and latency are taken into account. The number of nodes between two data points is defined as the number of "hops." The more hops, the longer the latency. Keeping latency and the number of hops down is important in applications such as live video and particularly in cases where PTZ cameras are used.

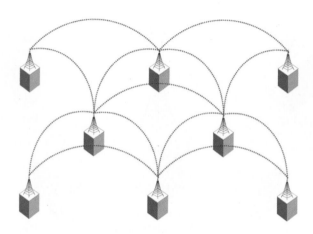

Figure 9.6 A wireless mesh network.

A wireless mesh network should be capable of managing itself. This means that if a node breaks down, the system will automatically set up a new path between two points. Wireless mesh networks have greatly improved over the past few years and are now a viable solution for video surveillance applications. Most solutions today are based on standard 802.11 technologies, with proprietary routing and security protocols.

9.3 802.11 WLAN Standards

The most common wireless standard for data networks is the 802.11 standard. It was published in 1997 by the IEEE and specifies a Media Access Control layer and different physical layers. The 802.11 standard describes two different transfer protocols as a solution for wireless communication: (1) the frequency hopping spread spectrum (FHSS) and (2) the direct sequence spread spectrum (DSSS), each of which enables data transfer at 1 and 2 Mbps. Both protocols use the 2.4 GHz frequency spectrum, which can be used worldwide without having to pay any license fees. In some regions, the signal strength is limited to 20 dBm (100 mW). Over the years, the 802.11 standard has been gradually improved and expanded. The following subsections outline the most relevant extensions of the standards.

9.3.1 The 802.11b Extension

The 802.11b extension was approved in 1999. It uses DSSS in the 2.4 GHz range, which results in data rates up to 11 Mbps. Until 2004, most WLAN products sold were based on 802.11b, and the products supported data rates of 1, 2, 5.5, and 11 Mbps, depending on the distance.

9.3.2 The 802.11a Extension

The 802.11a extension also was approved in 1999 and is based on the Orthogonal Frequency Division Multiplexing (OFDM) protocol. It operates in the 5 GHz frequency range. The 802.11a extension permits data rates of 6, 9, 12, 18, 24, 36, and 54 Mbps. Although higher data rates are achieved, there are also some downsides with 802.11a.

One of the main issues is that the 5 GHz frequency range is not available for use in parts of Europe where it is allocated for military radar systems. Consequently, certain adaptations were requested. Those adaptations include Transmit Power Control (TPC) and Dynamic Frequency Selection (DFS). TPC regulates the transmission power required for error-free data transfer so that too much coverage is prevented. DFS is designed to enable an automatic channel change if an external carrier is discovered on the channel used — which can be caused, for example, by a radar system. Both of these functions are specified in the 802.11h extension, which was approved in 2003. For this reason, it is important in Europe that 5 GHz WLAN components conform to 802.11a/h.

9.3.3 The 802.11g Extension

The 802.11g extension was approved in 2003 and operates in the 2.4 GHz range using the OFDM protocol. It provides data rates of 6, 9, 12, 18, 24, 36, and 54 Mbps. Because no comparable hurdles to the 5 GHz range exist in the 2.4 GHz range, the 802.11g solution quickly has established itself worldwide as the new standard, replacing 802.11b. Today, WLAN products are usually 802.11b/g compliant. They use the 2.4 GHz frequency and provide data rates up to 11 Mbps when DSSS is used and up to 54 Mbps when OFDM is used.

9.3.4 The 802.11n Extension

The IEEE is currently working on the 802.11n extension, which is expected to be approved during 2008. The 802.11n extension will enable data rates of up to 600 Mbps. These high rates are achieved with the Multiple Input Multiple Output (MIMO) protocol, where multiple antennas and spread paths for the electromagnetic waves are used to provide several transfer routes that can be used in parallel to make the high rates possible.

9.3.5 The 802.11s Extension

Another method of setting up a WLAN is referred to as a mesh network, which is specified in the yet unapproved 802.11s extension. All mesh networks available today use proprietary protocols, which means there is no compatibility between nodes from different manufacturers — an issue that the 802.11s standard is trying to address.

9.4 The Basics of 802.11 Networks

Different types of networks can be set up using the 802.11 technology. It can be as simple as a network camera and a laptop placed close to each other and directly exchanging data or as complex as several devices exchanging data between a number of nodes distributed across a wide area. The following subsections describe the basic 802.11 topologies, the frequencies in which they operate, and what channels mean.

9.4.1 Ad Hoc Network

The simplest type of 802.11 network consists of two nodes equipped with an appropriate interface. Each node with a WLAN interface forms a wireless cell, also called basic service set (BSS), in which other nodes can be located. If one node is located within the range of the first node and if the WLAN interface on these nodes functions using the same channel, data can be exchanged directly between the two nodes.

Ad hoc networks are the basic form of WLAN (wireless local area network) operation. To use this form of operation, the ad hoc mode must be selected as the operating mode on the WLAN interface. The range of coverage for an ad hoc network in a building is between 30 and 50 meters (98 and 164 feet) and in the open, up to 300 meters (984 feet). Figure 9.7 shows a network camera with built-in 802.11b/g and set to work in ad hoc mode.

Figure 9.7 A network camera with built-in 802.11b/g can be set to work in ad hoc mode. This is, however, not very common and not recommended.

9.4.2 Infrastructure Network

If larger areas need coverage with a WLAN, several BSSs can be connected and consolidated into a joint network. The individual BSSs are networked using what is known as a distribution system (DS). In most cases, the DS is a wired Ethernet network that connects the BSSs. Theoretically, a WLAN can reach any size by installing multiple BSSs and networking them with a DS. This is called an extended service set (ESS).

An access point has an Ethernet interface and at least one WLAN interface. When covering a large area, several access points can be used, whereby the individual access points must be positioned so that neighboring wireless cells slightly overlap one another to guarantee seamless transmission, as shown in Figure 9.8. In such a case, it is important that the wireless cells in the immediate vicinity function using different channels to avoid interference. After fulfilling these requirements, a user with

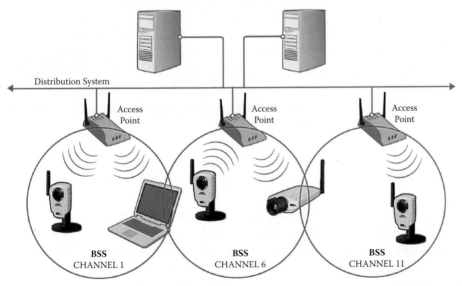

Figure 9.8 To provide appropriate coverage, the different access points' areas of coverage should overlap slightly.

a wireless device can move around the WLAN's complete coverage area without losing connection to the network or suffering performance losses.

Moving between wireless cells is known as roaming. The access points that form a network are identified by the Service Set Identifier (SSID). The SSID must be configured at each access point, and the nodes detect which access points belong to which network using the SSID, that is, which access points a node can associate with while roaming. If a device such as a network camera is to be integrated into an infrastructure network, the infrastructure operating mode, as well as the SSID, should be selected in order to connect to the right WLAN network.

9.4.3 802.11 Frequencies

As previously discussed, an 802.11 WLAN operates in the 2.4 or 5 GHz range. Both frequencies have their advantages and disadvantages. Networks that operate in the 2.4 GHz frequency are, by far, the most common. One of the major disadvantages with the 2.4 GHz frequency is that many other wireless technologies, such as Bluetooth, also use the same frequency. This can cause interference, with reduced data rates as a result.

The 5 GHz frequency range is not widely used. Although less interference occurs — which is a big advantage — there are disadvantages. As cited previously in the 802.11a subsection (Section 9.3.2), the frequency is used for radar systems in some European countries. Another disadvantage is that the higher the frequency used, the shorter the signal's reach. This is because the damping of electromagnetic signals increases as a function of the increased frequency. This can be particularly noticeable when signals have to penetrate barriers such as walls or furniture. Consequently, many more access points are required for transmissions in the 5 GHz range than in the 2.4 GHz range, which is one reason why 2.4 GHz is more widely used today.

9.4.4 Channels

The 2.4 GHz range is divided into 13 channels in Europe and 11 channels in the rest of the world. The channels are divided in such a way that neighboring channels overlap. As a result, only three independent wireless channels can be used in one wireless cell, for example, channels 1, 7, and channels 13, or 1, 6, and 11. This restriction should be taken into account when setting up a WLAN in the 2.4 GHz range, as the channels must be divided between the access points.

This effectively limits the number of cameras that can be used in a particular geographic area. The precise number of cameras that can be used is related to the average amount of bandwidth each will use. For example, if each camera uses 5 Mbps, approximately four cameras can be effectively connected per access point. With a maximum of three access points in an area, the limit is 12 cameras in the area. If this is too limited, consider using 802.11a, as the individual channels do not overlap when using the 5 GHz frequency.

9.5 WLAN Security

If data is transferred via a wireless link, it is important to take into account that a fixed boundary that can be found in a wired LAN does not exist and, in theory, anyone within range of a WLAN can intercept the data transferred and attack the network. Consequently, security becomes even more important in preventing unauthorized access to the data transferred and to the network. This section discusses some of the most common security technologies in wireless networking.

9.5.1 Wired Equivalent Privacy (WEP)

The basic 802.11 standard includes Wired Equivalent Privacy (WEP), a security protocol intended to provide a level of security comparable to that available for data transferred via cable. WEP is used to encrypt and authorize data. Encoding is based on a symmetrical encryption protocol, which is based on RC4 Stream Encryption from RSA. All nodes and access points that exchange data must have the same secret WEP key so that the data can be encrypted and decrypted. Depending on the WEP version, various key lengths are used. In the case of WEP40, the key length is 40 bits, and in the case of WEP128, the key length is 128 bits.

WEP uses a static key (i.e., it always remains the same) and must be entered manually at each node. To stop identical plaintext data from resulting in identically encoded data, the effective key is changed in each frame. This is achieved by extending the WEP with a variable, a 24-bit initialization vector (IV) set by the transmitting node every time a frame is sent. The receiving node also needs the IV, of course, so that it can decrypt the data again. To do this, the IV is transferred as plaintext in each frame.

Unfortunately, security loopholes have occurred because of the implementation of WEP. Today, there is a variety of utilities freely available on the Web that can be used to crack what is meant to be a secret WEP key. Consequently, it is no longer possible to consider WEP as adequate security. The IEEE is aware of this problem and introduced a new security protocol with the 802.11i extension in 2004. This includes the Temporary Key Integrity Protocol (TKIP), the use of the Advanced Encryption Standard (AES), and the authentication protocol to the 802.1X standard, which are described in more detail below.

9.5.2 Temporary Key Integrity Protocol (TKIP)

The TKIP (Temporary Key Integrity Protocol) was introduced to reliably deliver security via firmware or driver updates to WLAN products already installed. TKIP addresses the security loopholes in WEP and is a software program implemented on top of WEP. TKIP generates a new key for each session and works with an extended IV. Both measures ensure that the stream of encryption bits used does not repeat itself.

9.5.3 Advanced Encryption Standard (AES)

The AES (Advanced Encryption Standard) is an encryption algorithm recommended by the National Institute of Standards and Technology (NIST) and is based on what is referred to as the Rijndael Algorithm. Its high level of security and efficiency characterizes this algorithm. Until today, no crypto-analytical attack has the ability to crack the AES key. AES is implemented in the hardware, that is, the WLAN chip. A key length of 128 bits is used for AES in WLAN applications, with the user data being encrypted in blocks.

9.5.4 Pre-Shared Key (PSK)

TKIP and AES require keys from which a session key is derived. The Pre-Shared Key (PSK) protocol or authentication to 802.1X can be used to derive a session key. PSK is simpler than authentication to 802.1X. When using PSK, the key derives from a passphrase. The passphrase has a length of between 8 and 63 characters and must be entered manually into each node and access point. The PSK protocol is secure when it is based on a meaningless string of letters, numbers, and special characters such as 3aRs5%3?&d48fgH67, so that what is referred to as a "dictionary attack" cannot take place.

9.5.5 802.1X

Authentication to 802.1X can be used as an alternative to the PSK. In 802.1X, the individual nodes must identify themselves to a RADIUS (Remote Authentication Dial-In User Service) server before they can transfer data via a network. If the authentication is successful, a "pair-wise master key" is generated, from which the session key is derived. The use of 802.1X provides the advantage of being a central method, which is simple to administer. A RADIUS server is required, something normally found only in larger networks. Section 10.8.3 provides additional information about 802.1X.

9.5.6 Wi-Fi Protected Access (WPA2)

When the security risks of WEP became apparent, the network industry, under the sponsorship of the WiFi Alliance, had to react. The alliance defined Wi-Fi Protected Access (WPA) as the new security standard in 2003. In WPA, TKIP is used as the encryption technology. WPA2 started to be used in 2004, which corresponds to the AES implementation of 802.11i.

Depending on the type of security employed, different parameters must be set in a wireless network camera (Figure 9.9).

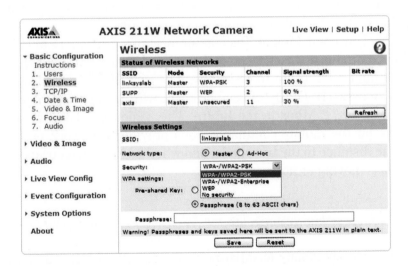

Figure 9.9 Depending on the type of security used, different parameters must be set in a wireless network camera.

9.6 Other Wireless Solutions

In addition to the 802.11 WLAN standards, there are many other technologies available for wireless transfer of data. Some of them are described below.

9.6.1 Bluetooth

Bluetooth was originally a wireless solution that enabled a data transfer rate of 1 Mbps in the 2.4 GHz frequency range over short distances. The original key application for Bluetooth was to wirelessly link peripherals over a distance of up to 10 meters (33 feet). The Bluetooth standard has been continually extended over the past few years so that data rates of up to 3 Mbps are possible today and distances of up to 100 meters can be bridged. Bluetooth is not commonly used in video surveillance applications.

9.6.2 Universal Mobile Telecommunications System (UMTS)

Wireless cellular phone networks are providing increasing abilities to transmit data. Whereas the bandwidth for data transfer was only 19.2 Kbps a couple of years ago, more and more cellular networks today use technologies that provide hundreds of kilobits per second. Soon megabits per second also will be available. This makes cellular networks increasingly interesting for video surveillance applications.

The Universal Mobile Telecommunications System (UMTS) is a mobile communications standard of the third generation (3G) that can achieve data rates of up to 384 kbps. The new High-Speed Downlink Packet Access (HSDPA) transfer protocol will be able to achieve data rates of up to 7.2 Mbps in the future. The achievable data rate always depends on the mobile communications provider's network expansion level, the distance to the base node, and whether the person transferring the data is moving or not moving.

9.6.3 Wireless Interoperability for Microwave Access (WiMAX)

Wireless Interoperability for Microwave Access (WiMAX) is a new long-distance mobile communications technology specified in the IEEE-802.16 standard. The target market is metropolitan area network (MAN) applications. With data rates of up to 70 Mbps, WiMAX is expected to compete with

UMTS. The achievable data rate will depend on the distance. It is expected that mobile communications providers in the future will offer WiMAX.

9.6.4 Proprietary Wireless Solutions

There are several proprietary wireless solutions available on the market. Some are applicable to video surveillance. There are many benefits with proprietary solutions, such as the use of different frequencies — some in the 900-MHz range, which gives superior distance. Proprietary solutions also normally have the ability to provide increased security. The downside is that the products are normally only compatible with the products of a single manufacturer, and the frequency used might be allowed only in some regions of the world.

9.7 Performance of Wireless Networks

The process of transferring data wirelessly is much more difficult than it is via cable. Consequently, a certain amount of data overhead for management, error recognition, and security is needed. In the case of wired Ethernet, the overhead is relatively small; that is, the net data rate is quite close to the gross data rate.

The net data rate for WLAN often amounts to considerably less than the gross data rate, often about 50 percent of the gross data rate specified. Furthermore, a wireless cell always must be considered as a shared medium. As a consequence, the bandwidth achievable inside a specific wireless cell depends on the number of nodes located in the cell.

When designing a wireless network for a video surveillance application and calculating the required bandwidth, the above must be taken into consideration to ensure that the desired frame rate is achievable.

9.8 Best Practices

For applications where a fixed data connection is not available and is too costly to install, a wireless solution may be the only solution. Before selecting the technology and implementing a wireless network, there are some important things to consider:

- *How many nodes will be networked?* If the answer is just two, then a point-to-point connection may be appropriate, and several proprietary as well as standard solutions are available. If several nodes will be networked, a multipoint or mesh network solution should be used.

- *What is the distance to be bridged?* The higher the frequency, the shorter the distance a certain ERP can cover. For very long distances, solutions in the 900-MHz range may be appropriate.
- *Are there any obstacles in the area of the wireless network?* A standard 802.11b/g network may have a reach of 30 to 100 meters indoors but up to 300 meters outdoors. Buildings, forests, and other obstacles will damper the wireless signal and limit its reach. The fewer the obstacles, the longer the distance that can be bridged.
- *What bandwidth is required?* Most wireless solutions specify the maximum bandwidth. The real bandwidth depends on the distance and obstacles and may be half or even only 10 percent of the maximum.
- *What is the level of risk for interference?* It is important to consider if other wireless networks will be installed in the area. Check existing equipment in the area with a spectrometer, which will be able to indicate the frequencies in use.
- *How secure must the connection be?* Wireless means sending data in the open air. If used for video surveillance, it is important to secure the data using the appropriate security protocols, such as 802.1X. Remember that adding security protocols normally decreases the performance.

A wireless network is a cable replacement technology. Power is still required at both ends. Power over Ethernet (PoE) reduces the need for a power cable but retains the data cable. For most indoor video surveillance applications, using PoE is both more cost efficient and secure. For more information on Power over Ethernet, see Section 8.6.

Chapter 10

Networking Technologies

Networking technology has seen tremendous development over the past 15 years and today is used by all on a daily basis. Although the technology is quite complex, it is something that users fortunately do not experience. When a laptop is turned on, a complex series of networking technologies is automatically initiated to ensure that the laptop gets an IP address and starts to communicate securely over a network. The Internet Protocol is the common denominator in network technology, and although it was originally designed for military communications, it also is used today in small home networks, enterprise LANs, and the Internet for applications such as e-mail, Web browsing, telephony, and network video.

Many different protocols are used when data is securely transferred from one networked device to another. The best way to understand how the different protocols interact is by understanding the OSI communication model, which is explained in the first section of this chapter. The IP protocol is discussed extensively in the middle of the chapter. The chapter ends with a section on network security, a topic that is increasingly important as networks and their use become more widespread.

10.1 OSI Reference Model

Data communication between open systems is described using the Open Systems Interconnection (OSI) Reference Model, which is composed of seven layers (Figure 10.1). Each layer provides specific services and makes the results available to the next layer. To provide a service, each layer

Layer 7-Application	Data
Layer 6-Presentation	Data
Layer 5-Session	Data
Layer 4-Transport	Segments
Layer 3-Network	Packets
Layer 2-Data Link	Frames
Layer 1-Physical	Bits

Figure 10.1 The OSI Reference Model includes seven layers. Each layer performs a certain service.

utilizes the services of the layer immediately below it. Communication between layers occurs by means of specific interfaces. Each layer must follow certain rules known as protocols to perform its services. It is important to note that the OSI is not a protocol itself but, rather, a model used to understand the function of protocols.

Communication between two systems always occurs on the same layer and is known as virtual communication. Virtual communication can take place only if the same protocols are implemented in the corresponding layer of another system. For example, one system passes the data to be transferred physically downward to the lowest vertical layer and then transfers it horizontally to the other system, where the data is then passed vertically upward. In this way, the data reaches the corresponding layer on the other system and communication takes place.

10.1.1 Layer 1: The Physical Layer

The Physical layer is the lowest layer and provides services that support the transmission of data as a bitstream over the relevant medium such as a wired or wireless transmission link. This layer describes the transmission medium and its physical characteristics, as well as the mechanical and electrical means that permit a physical connection to the transmission medium. That is, this layer defines the forms of electrical signals, optical signals, or electromagnetic waves, as well as the connectors and jacks needed for a connection to a network cable.

Examples of the protocols and standards that operate in this layer include IEEE 802.3 (Ethernet) and IEEE 802.5 (Token Ring).

10.1.2 Layer 2: The Data-Link Layer

The Data-Link layer provides data transmission and controls access to the transmission medium. This is achieved by combining data into units known as frames. The frames are provided with a checksum, which is used by the recipient to detect possible transmission errors. According to the IEEE, the second layer is subdivided into two sublayers, with the upper range corresponding to the Logical Link Control (LLC) and the lower part corresponding to the Media Access Control (MAC). LLC is used in the same way by all IEEE network technologies and simplifies data exchange. MAC controls access to the transmission medium and depends on the network technology used.

Typical protocols and standards are IEEE 802.2 (LLC), IEEE 802.3 (Ethernet MAC), and 802.11 (WLAN MAC).

10.1.3 Layer 3: The Network Layer

The third layer performs the actual data transfer between systems by performing the routing and forwarding of data packets between the systems. The most important tasks of the Network layer include the creation and administration of routing tables. The Network layer provides options for communicating beyond network boundaries. Destination and source addresses are assigned to data in this layer and are used as a basis for targeted routing. Addressing in the third layer is independent of addressing at lower levels, which means that routing can extend to multiple logically structured networks.

Examples of protocols that operate in this layer are the Internet Protocol (IP) and Routing Information Protocol (RIP).

10.1.4 Layer 4: The Transport Layer

The function of the Transport layer is to provide a reliable data transfer service to Layer 5 and above. The Transport layer controls the reliability of a wired or wireless link through flow control and error control. Some protocols that perform in this layer are state and connection oriented, which means that they can keep track of segments and retransmit those that fail.

Examples of protocols in this layer are the Transmission Control Protocol (TCP) and User Datagram Protocol (UDP).

10.1.5 Layer 5: The Session Layer

The Session layer provides an application-oriented service and takes care of the process communication between two systems. Process communication begins with the establishment of a session, which provides the basis for a virtual connection between two remote systems.

Examples of protocols that provide Layer 5 functions include Remote Procedure Call (RPC) and Network File System (NFS).

10.1.6 Layer 6: The Presentation Layer

The Presentation layer converts system-dependent data formats, such as ASCII, into an independent format and thus permits syntactically correct data exchange between different systems. The tasks of this layer also include data compression and encryption. The Presentation layer ensures that data sent by the Application layer of a system can be read by the Application layer of another system.

Examples of protocols that provide Layer 6 functions include X.216 (Presentation Service) and X.226 (Connection-Oriented Presentation Protocol).

10.1.7 Layer 7: The Application Layer

The Application layer is the highest layer of the OSI model. It makes a variety of functions, such as Web, file, and e-mail transfers, available to applications. The actual applications, such as a Web browser or Microsoft Outlook, lie above this layer and are not covered by the OSI model.

Typical protocols in this layer are the File Transfer Protocol (FTP), Simple Mail Transfer Protocol (SMTP), and HyperText Transfer Protocol (HTTP).

10.2 The TCP/IP Reference Model

The TCP/IP Reference Model can be used to understand protocols and how communication takes place. In this model, the different protocols fall

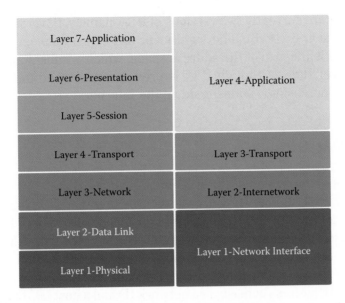

Figure 10.2 The TCP/IP Reference Model can be split up into four layers (right), comparable to the seven in the OSI model (left).

into four different layers, which correspond to the seven layers in the OSI model (Figure 10.2).

10.2.1 The Internet Protocol (IP)

The Internet Protocol (IP) is a Layer 3 protocol in the OSI model and a Layer 2 protocol in the TCP/IP Reference Model. It is the key protocol in most networked applications today.

Every server on the Internet must have its own individual public IP address. For a server to operate on the Internet, an IP address must be requested from an Internet Network Information Center (InterNIC) or allocated by the Internet service provider (ISP). An ISP can allocate either a dynamic IP address, which can change during a session, or a static address, which normally comes with a monthly fee.

IP offers a connectionless service that splits data into IP datagrams before they are transmitted. "Connectionless" means that IP does not guarantee whether and in what sequence IP datagrams will arrive at a recipient. Because IP operates on a connectionless basis, individual IP datagrams can be routed to a recipient in different ways. Guaranteed data transmission and reassembly into the correct sequence are performed by protocols in the Transport layer of the OSI and TCP/IP models.

There are currently two versions of IP — namely, IP version 4 (IPv4) and IP version 6 (IPv6). The significant difference between these two IP versions is the size of the address space, which is substantially larger in the newer IPv6. Although IPv4 is predominantly used today, IPv6 is becoming increasingly widespread on the Internet.

10.2.2 IPv4 Addresses

An IPv4 address has a length of 32 bits, meaning that 2^{32}, or 4.3 billion (4,294,967,296), unique IP addresses can be assigned. To simplify readability, IP addresses are grouped into four blocks of 1 byte (8 bits) each. The individual blocks (referred to as octets) are separated by a point, and each block represents a decimal value between 0 and 255 (e.g., 85.235.16.37). This is referred to as dot-decimal notation.

Each IP address consists of a network ID and a host ID. The network ID represents the logical IP network in which the host resides. The host ID represents the host itself. All devices with the same network ID portion of an IP address will reside on the same network segment. In this case, the hosts can communicate directly with each other without the use of a router to forward the traffic to the correct network.

IP addresses can be classified into five groups referred to as classes (Table 10.1). IP addresses are grouped into the various classes based on the number range of the first octet of the IP address. Class C addresses are used in most applications. The groups define how many network and host addresses are available within the class. IP addresses used in LANs and WANs come from classes A, B, and C; classes D and E are for multicast and experimental uses.

Note, however, that the class-based addressing model of IPv4 is not commonly used today. Instead, Classless InterDomain Routing (CIDR) is used, which permits a more efficient use of scarce IPv4 addresses.

Table 10.1 Classes of IP Addresses

Class	Value Range of First Byte	Bytes for Net ID	Number of Networks	Bytes for Host ID	Number of Hosts
A	1–126	1	126	3	16,777,214
B	128–191	2	16,384	2	65,534
C	192–223	3	2,097,152	1	254
D	224–239	Multicast addresses	N/A	N/A	N/A
E	240–254	Reserved	N/A	N/A	N/A

Table 10.2 Explanation of the Permitted Values for the Subnet Mask Bits

128	64	32	16	8	4	2	1	—
1	0	0	0	0	0	0	0	128
1	1	0	0	0	0	0	0	192
1	1	1	0	0	0	0	0	224
1	1	1	1	0	0	0	0	240
1	1	1	1	1	0	0	0	248
1	1	1	1	1	1	0	0	252
1	1	1	1	1	1	1	0	254
1	1	1	1	1	1	1	1	255

10.2.3 Subnets

An IPv4 address is divided into two parts: (1) the network ID and (2) the host ID. The subnet mask determines the blocks of an IPv4 address that define the network and host identifiers. A subnet mask has a length of 32 bits and also is represented in dot-decimal notation (e.g., 255.255.255.0).

When determining the network and host IDs within an IP address, the computer converts the IP address and subnet mask into binary. Table 10.2 shows the binary-to-decimal numbering conversions.

When viewed as a binary number, the bits of the subnet mask are a contiguous group of 1's followed by a contiguous group of 0's. The bits of an IP address that correspond to the group of 1's in the subnet mask represent the network ID. The bits of an IP address that correspond to the group of 0's in the subnet mask represent the host ID of an IP address. Only subnet masks with contiguous bits are allowed.

A permitted subnet mask thus corresponds, for example, to:

- 255.255.192.00 (11111111.11111111.11000000.00000000)
- 255.255.255.0 (11111111.11111111.11111111.00000000)

The following subnet mask contains non-contiguous bits and thus would not be permitted:

- 255.255.230.0 (11111111.11111111.11100110.00000000)

Table 10.3 provides some examples of IP addresses that are broken into the network and host ID. In Example 1, the subnet mask indicates that the first three blocks of digits in the IP address define the network ID and the last block, the host ID. In Example 2, the subnet mask indicates that

Table 10.3 Examples of IP Addresses Broken into Network ID and Host ID

	Example 1	Example 2
IP address	192.168.1.5	10.50.88.129
Subnet Mask	255.255.255.0	255.255.0.0
Network ID	192.168.1.0	10.50.0.0
Host ID	0.0.0.5	0.0.88.129

the first two blocks define the network ID, whereas the last two blocks define the host ID.

If a host wants to transmit data, it determines the network ID of the destination IP address by means of the subnet mask. If the destination IP address has the same network ID as the host that is transmitting the data, then the host sends the data directly to the destination.

If the network ID of the destination IP address is located in another subnet, then the host sends the data to the default gateway defined in the host. The default gateway is the IP address of a router that forwards the data to the correct network. An alternative notation based on IPv4-CIDR (Classless Inter-Domain Routing) exists for representing the IP address and subnet mask and uses a suffix to indicate the length of the subnet mask (number of 1's). The number of bits used for the subnet mask is appended to the IPv4 address as a decimal number followed by "/". For example, 192.168.12.23/24 corresponds to the IP address 192.168.12.23 with the subnet mask 255.255.255.0.

10.2.4 NAT

Three address ranges are defined and used exclusively for private purposes. These private IP addresses are 10.0.0.0 to 10.255.255.255, 172.16.0.0 to 172.31.255.255, and 192.168.0.0 to 192.168.255.255. The addresses can be used only on private networks and are not allowed to be forwarded through a router to the Internet. If Internet connectivity is required, it can be solved using a technique called Network Address Translation (NAT) protocol, which operates on Layers 3 and 4 of the OSI model.

In such cases, IP datagrams are forwarded via a router that supports the NAT protocol, which transparently readdresses a private IP address into a public IP address without the sending host's knowledge. The public IP address is the address allocated by the Internet provider. Larger private networks with multiple servers also can use the same readdress-

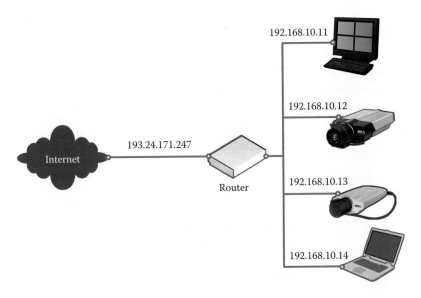

192.168.10.11

192.168.10.12

193.24.171.247

Internet

Router

192.168.10.13

192.168.10.14

Figure 10.3 Multiple servers can connect to the Internet via a unique public IP address using a router and NAT.

ing technique. The router manages the connection between the private network and the Internet (Figure 10.3).

10.2.5 Services and Port Numbers

Data that will be sent from one system to another must be associated with a particular service or application so that the receiving server will know how to process the incoming data. This is done by associating the data with a service or process that is defined by, or mapped to, a particular port number on a server. For example, a service that runs on a Web server typically is mapped to port 80 on a computer (Figure 10.4).

A port number has a length of 16 bits, that is, values from 0 to 65535. Certain applications use port numbers that are preassigned to them by the Internet Assigned Numbers Authority (IANA). These port numbers range between 0 and 1023 and are also known as "well-known registered ports." The ports between 1024 and 49151 are known as "registered ports." Manufacturers of applications can, on request, register ports from this range for their own protocols. The remaining ports, ranging from 49152 to 65535, are called "private ports." They can be used flexibly because they are not registered and are therefore not assigned to any application.

Many network video products permit reconfiguration of the port numbers of individual services. For example, the port number of the Web

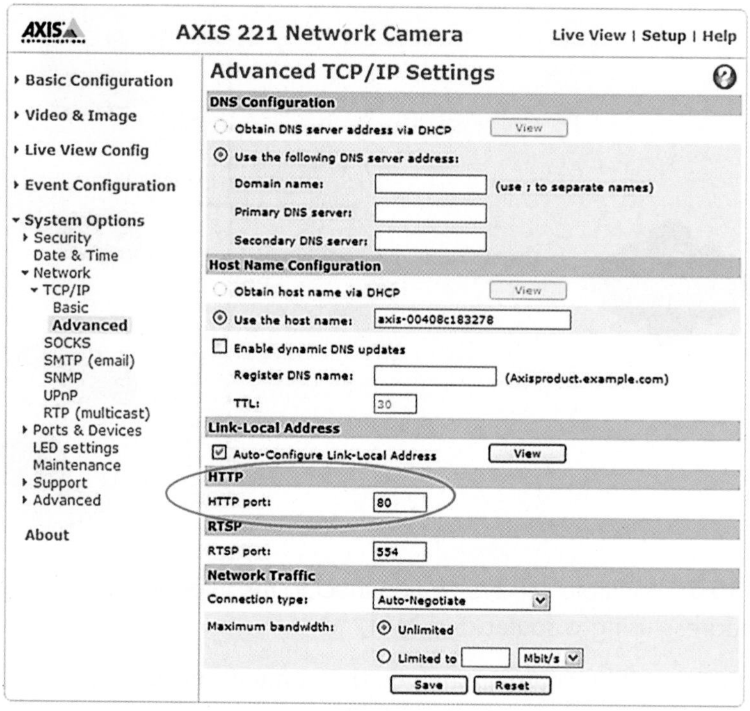

Figure 10.4 In many network cameras, the port number can be set. Normally, the port is set to port 80 if the network camera is accessed as a Web service via HTTP.

server service on network cameras can be changed from port 80 to private ports. Using private ports also means that it will be more complex to access a network camera via a Web browser, which may be desirable in some applications.

10.2.6 Port Forwarding

To configure access from the Internet to cameras located on a private LAN, one can use a technique called "port forwarding." Port forwarding is available in most routers today.

Port forwarding works as follows (Figure 10.5). Incoming data packets reach the NAT-enabled router via a public IP address. The router is configured to forward any data coming into a predefined port number to a specific host on the private network side of the router. The router then replaces the address of the incoming packet with a private IP address. To a receiving client, it looks like the source of the packets originated from the router. The reverse happens with outgoing data packets. The

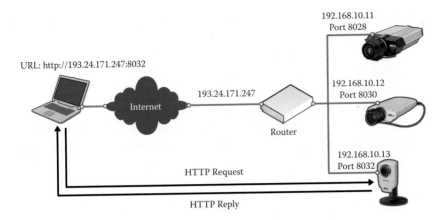

Figure 10.5 With port forwarding, network cameras on a local network can be individually addressed over the Internet.

router replaces the private IP address of the source device with a public IP address before sending out the data over the Internet.

To access one of the network cameras over the Internet, the public IP address of the router should be used together with the corresponding port number of the device on the private network. In the system illustrated in Figure 10.5, typing the URL http://212.134.245.45:8032 into a Web browser would give the user access to a device with a private IP address of 192.168.10.13 Port 8032.

10.2.7 IPv6

Using protocols such as NAT, the shortage of IP addresses in IPv4 has thus far been managed. In the future, however, the number of IPv4 addresses will no longer be sufficient, and the use of IPv6 therefore will become more widespread.

The major advantages of IPv6, apart from the availability of a huge number of IP addresses, include IP auto-configuration based on the MAC address, renumbering to simplify switching entire corporate networks between providers, faster routing, point-to-point encryption according to IPSec, and connectivity using the same address in changing networks (Mobile IPv6).

10.2.7.1 IPv6 Addresses

An IPv6 address has a length of 128 bits, meaning 2^{128}, that is, approximately 340.28 sextillion (3.4×10^{38}), addresses. An IPv6 address is written in hexadecimal notation with colons subdividing the address into eight blocks of 16 bits each. An example of an IPv6 address is:

2001:0da8:65b4:05d3:1315:7c1f:0461:7847

To simplify how the address is represented, two consecutive colons can replace one or more 16-bit groups with the value 0000. However, the resulting address can contain only two consecutive colons once. The address 2002:0da8::1315:7c7a is therefore equivalent to:

2002:0da8:0000:0000:0000:0000:1315:7c7a

Leading zeroes in a 16-bit group also can be omitted; for example, 2002:0da8::0017:000c can be written as 2002:da8::17:c.

Address ranges in IPv6 are indicated by prefixes; subnets also are determined by the prefix. The prefix length (number of bits) is appended to the IPv6 address as a decimal number followed by "/" (forward slash). Subnet masks as used in IPv4 no longer exist in IPv6; instead, a similar notation to IPv4-CIDR is used.

In IPv6, the first 64 bits of the address are usually intended for network addressing, and the last 64 bits are used for host addressing. For example, if a system has the following IPv6 address:

2002:0da8:67f3:08a4:1511:aa56:0361:7a4f

then the system comes from the subnet:

2002:0da8:67f3:08a4::/64

IPv6 enables a device to automatically configure its IP address using the MAC address. In these cases, the prefix — being the first 64 bits — is always the same; fe80 and the remaining 48 bits correspond to zeroes (fe80:0000:0000:0000). For the remaining 64 bits (suffix) of the IPv6 address, the MAC address of the system is converted into the Extended Unique Identifier-64 (EUI-64) numbering system. The result would be, for example:

fe80::1511:aa56:0361:7a4f

The MAC-based address permits a networked device to communicate on the local network. However, for communication over the Internet, the first 64 bits of the IPv6 address must be adapted to the network address

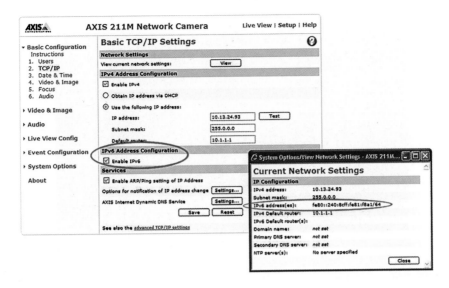

Figure 10.6 Many network cameras have support for IPv6, which will become increasingly important in network video applications.

of the router, which has been allocated to the router by the ISP. To do this, a system sends the router a corresponding host request and receives the necessary prefix of the public address block and additional information from the router. Using this information, a system then can create the IPv6 address from the prefix and its suffix (EUI-64 address). Services such as DHCP for IP address allocation and tasks such as the manual configuration of IP addresses are therefore no longer required in IPv6.

The IPv6 address is enclosed in square brackets in a URL. An example of a correct URL is:

http://[2002:0da8:67f3:08a4:1511:aa56:0361:7a4f]/.

A specific port also can be addressed by changing the address as follows:

http://[2002:0da8:67f3:08a4:1511:aa56:0361:7a4f]:8081/.

See Figure 10.6.

10.3 Managing IP Addresses

In the early days of networking, most devices had a fixed IP address that was set and managed manually. As networks grew, so did the demand for

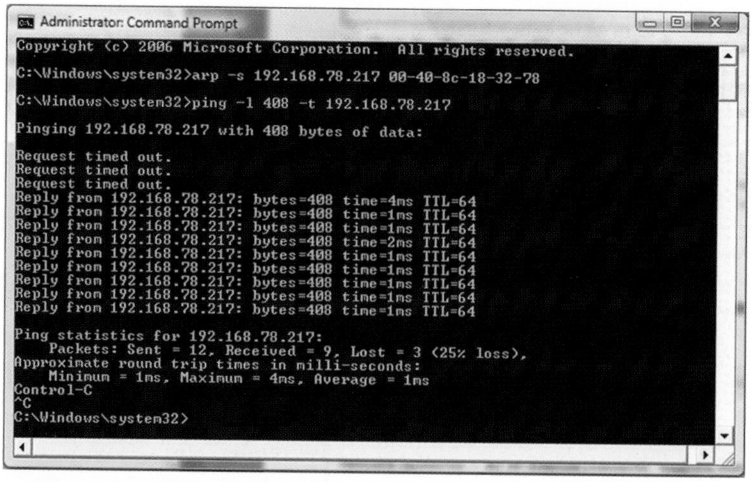

```
Administrator: Command Prompt                                    _ □ X
Copyright (c) 2006 Microsoft Corporation.  All rights reserved.

C:\Windows\system32>arp -s 192.168.78.217 00-40-8c-18-32-78

C:\Windows\system32>ping -l 408 -t 192.168.78.217

Pinging 192.168.78.217 with 408 bytes of data:

Request timed out.
Request timed out.
Request timed out.
Reply from 192.168.78.217: bytes=408 time=4ms TTL=64
Reply from 192.168.78.217: bytes=408 time=1ms TTL=64
Reply from 192.168.78.217: bytes=408 time=1ms TTL=64
Reply from 192.168.78.217: bytes=408 time=2ms TTL=64
Reply from 192.168.78.217: bytes=408 time=1ms TTL=64
Reply from 192.168.78.217: bytes=408 time=1ms TTL=64
Reply from 192.168.78.217: bytes=408 time=1ms TTL=64
Reply from 192.168.78.217: bytes=408 time=1ms TTL=64
Reply from 192.168.78.217: bytes=408 time=1ms TTL=64

Ping statistics for 192.168.78.217:
    Packets: Sent = 12, Received = 9, Lost = 3 (25% loss),
Approximate round trip times in milli-seconds:
    Minimum = 1ms, Maximum = 4ms, Average = 1ms
Control-C
^C
C:\Windows\system32>
```

Figure 10.7   Setting a static IP address using the ARP and PING commands.

techniques to manage all the networked devices to make the tasks of setting and tracking IP addresses as automated as possible.

The section below describes the different ways of setting and managing IP addresses. IP addressing of data packets and the use of Domain Name System are also explained.

10.3.1 Setting IP Addresses

Any device on an IP-based network must have a unique and appropriate IP address. Setting the IP address can be done in two different ways: (1) manually using a static address or (2) dynamically using DHCP.

10.3.1.1 Manual Address Allocation

Setting static IP addresses is labor intensive. Static addresses are normally used only in smaller systems or for a few devices on a network that need a static address for certain reasons. One way to set the IP address of a network camera is to use the ARP and commands, normally executed from the DOS prompt (Figure 10.7). When set, the correct subnet mask must be set before the network camera can be accessed from other subnets.

Many vendors offer different types of tools that help not only set IP addresses but also, more importantly, find and manage the devices on a network. In a network video system with potentially hundreds of network cameras, such tools are a necessity to effectively manage the system. (See Figure 10.8.)

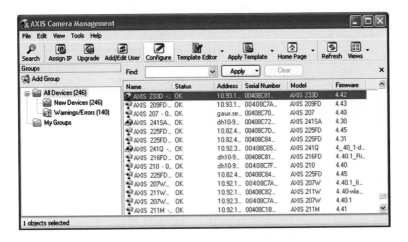

Figure 10.8 Most vendors provide tools that make it easier to manage the IP addresses of networked devices.

10.3.1.2 Dynamic Address Allocation

The Dynamic Host Configuration Protocol (DHCP) is the standard protocol used for automatic assignment and management of IP addresses. A DHCP server manages a pool of IP addresses, which it can assign dynamically to a DHCP client upon request. The DHCP server also can provide other IP configuration parameters to DHCP client computers such as a host name and a DNS (see Section 10.3.4) server address. A DHCP server is the central point of client configuration management; it removes the need to maintain individual network configurations for each network client.

When a DHCP client comes online, it sends a query requesting configuration from a DHCP server. The DHCP server replies with an IP address, subnet mask, and other client configuration parameters. In a typical configuration, the IP address provided by a DHCP server is "leased" to the client. Once half of the lease time has expired, the client will request to renew the lease. (See Figure 10.9.)

10.3.2 Configuration-Free Networking

In many systems, especially small ones, setting and managing IP addresses is complex and cumbersome. To address these issues, several techniques have been developed to simplify and automate IP addressing to the highest extent possible. The two most well-known techniques are UPnP/Zeroconfig and Bonjour.

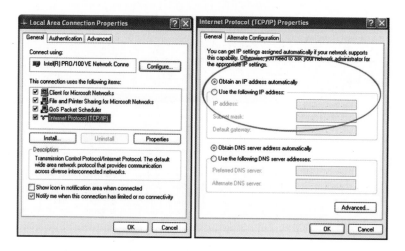

Figure 10.9 Any networked device, such as a PC, must have a unique IP address, which can be set dynamically via DHCP.

10.3.2.1 UPnP and Zeroconfig

Zeroconfig is a component of Universal Plug and Play (UPnP). Using UPnP, Microsoft operating systems such as Windows XP or Vista automatically can detect resources on a network, which means that a network camera will automatically be listed in My Network Places in the Windows operating system.

Zeroconfig is a procedure for configuration-free system networking in LANs. Zeroconfig, for example, can be used to connect two notebooks with a crossover cable. The notebooks independently allocate themselves an IP address and can exchange data without the need to configure the IP address or use a DHCP service.

With Zeroconfig, the networked devices attempt to independently allocate an IP address between 169.254.1.0 and 169.254.254.255. If a system wishes to configure an IP address, it simply selects an IP address from the range using a random number generator based on system-specific information, such as the MAC address. After selecting an appropriate IP address, the system must first check whether it is already in use by another networked device.

10.3.2.2 Bonjour

Bonjour is another protocol for the announcement and discovery of services in an IP network. This protocol is based on open standards and was introduced by Apple. Using Bonjour, services can be discovered independently, and necessary configurations can be performed automatically without any user intervention. Bonjour is based on the exchange of multicast DNS packets sent via the UDP port 5353 (224.0.0.251).

In addition to multicast DNS, Bonjour is based on DNS Service Discovery (DNS-SD). DNS-SD is an expansion of the Domain Name Service as used for domain management. Bonjour is comparable to UPnP and Zeroconfig.

Bonjour is appropriate for use in discovering network video products using Mac computers, but it can also be used as a discovery protocol for new devices in any network.

10.3.3 MAC and IP Address Resolution

IP addressing of data packets operates on the third OSI layer. Before data can be transported over a physical network, it must be packed into frames, which occurs in the Data-Link (second) layer of the OSI model. Addressing on the second layer uses MAC addresses, which means that in addition to knowing the destination IP address, the sending host must know the MAC address of the destination host so that the destination IP address can be associated with a MAC address. To find out the MAC address of a given destination host, the sending host sends out a request using a protocol called ARP.

10.3.3.1 Address Resolution Protocol (ARP)

The Address Resolution Protocol (ARP) is used to discover the MAC address of the destination host. An ARP request specifies the IP address of the destination host, in addition to the IP and MAC addresses of the sending host. When a switch receives an ARP request, it broadcasts the ARP request to all devices on the local segment. The devices then compare the destination address to their own IP addresses. The device with the corresponding IP address replies by disclosing its MAC address.

The requesting host then enters the MAC address it receives, together with the associated IP address, into its ARP cache. The ARP cache is a table that temporarily stores the mapping of addresses, usually between MAC and IP addresses, in the RAM memory of a host. Before sending out the data, the sending host checks whether it can resolve the necessary MAC address using its ARP cache. If this cannot be done, then it sends out an ARP request (see Figure 10.10).

When a destination host resides outside a sending host's LAN, the sending host must use the MAC address of its router instead because it is not able to find out the MAC address of the destination host (Figure 10.11).

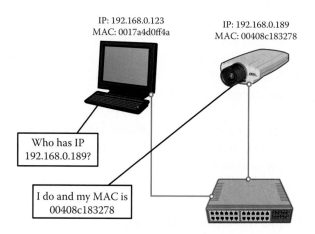

Figure 10.10   To send data over a local network, the ARP is used so that the sending host can properly address the data packet with the destination host's IP and MAC addresses.

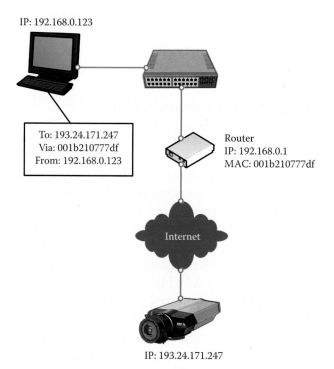

Figure 10.11   To send data to a destination host outside a local network, the sending host uses its router's MAC address, together with the destination host's IP address.

If the sending host does not know the MAC address of its router, an ARP request is sent to discover it. Thus, to properly address a data packet that will be sent to another LAN, the sending host uses the MAC address of its router, together with the IP address of the destination host.

10.3.4 Domain Name System (DNS)

The Domain Name System (DNS) converts domain names into their associated IP addresses and operates on the Transport layer. This can be compared with a telephone book, which links the name of a person with his phone number. The DNS-to-IP address mapping can be done either manually in a host file or automatically, which is more common.

DNS caters to the human ability to better remember names than numbers. For example, the domain name *company.com* is much easier to remember than its associated IP address *193.24.171.247*. DNS is used on a daily basis when a Web site is accessed by typing a domain name, such as www.company.com, into a Web browser's address line. DNS then converts this name into the IP address in the background, and communication between the systems takes place via the IP addresses.

10.3.4.1 Dynamic DNS

When a PC or network camera connects to the Internet, typically the Internet service provider will assign an unused IP address dynamically, via DHCP. This address can be used for only a short time, and several addresses can be used throughout the length of time a connection is maintained. In such scenarios, a Dynamic DNS (DynDNS) is used to keep track of a domain name's link to varying IPv4 addresses.

In Dynamic DNS, the host record on the DNS server is updated whenever the host's IP address is changed. Either the host itself or the DHCP server, depending on the host's capabilities and network configuration, sends this update.

Using Dynamic DNS, a network device's domain name — which never changes — can be used to access the device, regardless of the IP address currently assigned to it.

10.4 Data Transport

Internet Protocol (IP) is the most important protocol in the Network layer and the core protocol for any data communication. However, IP never appears alone but always together with protocols in the same layer or higher layers. In the Transport layer (Layer 4 in the OSI model), the most common protocols used are User Datagram Protocol (UDP) and Transmission Control Protocol (TCP). These protocols, along with the application protocols in Section 10.5, generally are referred to as the IP or TCP/IP protocol suite.

10.4.1 User Datagram Protocol (UDP)

The User Datagram Protocol (UDP) is a network protocol located on the fourth OSI layer and provides a connectionless transmission service. The fact that the protocol is connectionless means that no connection is established between the sender and receiver.

Messages are packetized into datagrams and transmitted. If the datagrams make it through the network, they are received by another application. If datagrams are lost, there is no retransmission strategy. Consequently, there is no in-order guarantee. A UDP receiver can deliver data to the application in the wrong order.

Ports are used in UDP to allocate data to the correct application in the destination system. To this end, the port number of the service that should receive the data is embedded in the UDP header. UDP also sends a checksum in its header as a reliable integrity check option. This enables the recipient to detect transmission errors.

The UDP does not apply any flow or congestion control to its sending strategy. If applications generate large amounts of data, the network can be flooded. A common scenario is that applications send more data than what can be sent through on a network. This leads to packet losses for its own data and possibly for other flows.

From a video surveillance perspective, UDP favors timely delivery of data over reliability but fails to provide congestion and burst control — two features that TCP provides. The UDP tends to be a better choice when trying to minimize delays and jitter. Using UDP may be preferable when transmitting data that requires low latency and that can tolerate some losses, such as with multimedia broadcast applications. On the other hand, when bandwidth is short or a firewall or NAT is in the path, TCP tends to work better.

10.4.2 Transmission Control Protocol (TCP)

The Transmission Control Protocol (TCP) is the most commonly used protocol for data transport, and when used with IP, the protocol often is referred to as TCP/IP. TCP divides data into TCP segments for data transmission, adding supplementary flow-control information in the TCP header. TCP provides a connection-oriented, reliable, and in-order delivery of data streams. In addition, it is responsive to network congestion. These characteristics make TCP suitable for applications such as file transfers or e-mail.

In contrast to UDP, TCP is a connection-oriented protocol. This means that it establishes a connection between two communicating applications before any data exchange takes place. The connection makes sure that data flows only between two hosts. This disables the possibility of using TCP for broadcasting or multicasting. TCP also provides transmission reliability: the recipient confirms the incoming data and, when necessary, the sender retransmits if no confirmation is received.

The TCP provides transmission and reception of a reliable and in-order stream of bytes for applications at upper layers. It does this by splitting the payload into sequence-numbered segments, transmitting those segments according to its protocol rules, and finally, at the receiver, verifying and reassembling the segments to reconstruct the original stream. As data is exchanged, receivers continuously acknowledge successful receipt of segments by issuing sequence numbers so that senders can retransmit lost data when needed. When segments are lost, a TCP receiver does not pass any out-of-order data to the applications. It waits for the retransmitted segment to arrive; that is, TCP introduces a delay in its effort to be reliable and in-order.

While exchanging data, TCP continuously applies flow and congestion control to its sending strategy. For example, when a receiver is slower than the sender, TCP's flow control forces the sender to slow down. Similarly, when the network path is congested, TCP's congestion control forces the sender to slow down. These are important features that avoid congestion collapses, excessive packet loss, and unfairness between flows.

From a video surveillance perspective, conventional wisdom indicates that TCP is not suitable for real-time traffic because it favors reliability over a timely delivery of data.

Because of this, many real-time protocols tend to use UDP and only fall back on TCP when absolutely necessary (e.g., due to bandwidth, firewall, or NAT issues in the path). On the other hand, for video applications where delays are not critical, such as for video storage, TCP is commonly used.

10.5 Application Layer Protocols

On the Application layer, which is the highest level of both the OSI model and the TCP/IP model, different protocols are required for data exchange between a network video system and the user to display, for example, the menus of a network camera and to view videos. The most common protocols are HTTP, FTP, and RTP, which are explained below.

10.5.1 HyperText Transfer Protocol (HTTP)

The HyperText Transfer Protocol (HTTP) is used primarily to load the text and images from a Web site to a Web browser. HTTP is a stateless protocol, meaning that a connection is not maintained between systems once data has successfully transmitted. A new connection must be established for further data transmissions.

Network video systems provide an HTTP server service that permits access to the systems via Web browsers to download configurations or live images.

10.5.2 File Transfer Protocol (FTP)

The File Transfer Protocol (FTP) is a network protocol for data transmission via TCP/IP. It primarily is used to transmit files from a server to a client (download) or from a client to a server (upload). FTP also can be used to create and select directories and rename or delete directories and files.

There are two modes for establishing FTP connections: active mode and passive mode. In active mode, the FTP server establishes a connection to the client following a request from the client, whereas in passive mode the client establishes the connection to the server. Passive mode is used if the server cannot reach a client. This is the case, for example, if the client is located behind a router that converts the client's address by means of NAT or if a firewall protects the local network against external access.

Network cameras or video encoders can use FTP to transmit JPEG images or MPEG-4 video sequences to an FTP server for archiving purposes. In such a case, the network camera acts as an FTP client and establishes an event-based connection to the FTP server. It then transmits multiple JPEG images to the server and stores them to a specific directory using different file names.

10.5.3 Simple Network Management Protocol (SNMP)

The Simple Network Management Protocol (SNMP) represents a set of protocols for managing complex network infrastructures. It can be used to remotely monitor and manage networked equipment such as switches, routers, and network cameras. Many network cameras have support for SNMP, which means that they can be managed by tools such as Openview from Hewlett-Packard and Tivoli from IBM. The latest version of SNMP is version 3.

10.5.4 Simple Mail Transfer Protocol (SMTP)

The Simple Mail Transfer Protocol (SMTP) is the *defacto* standard for transferring e-mail over the Internet. Although it is not normally relevant in network video, many network cameras and encoders have support for SMTP to enable the possibility to e-mail alerts that notify, for example, that motion was detected and even attach a snapshot or a video clip.

10.5.5 Real-time Transport Protocol (RTP)

Real-time Transport Protocol (RTP) permits the transfer of real-time data between system endpoints. RTP is a packet-based protocol that is generally transmitted via UDP. The RTP services include identification of the transmitted user data and its sources and the allocation of sequential numbers and timestamps to the data packets. With this information, the recipient is capable of reassembling the individual data packets in the correct sequence.

Video compressed using MPEG-4 is often transmitted via RTP. In conjunction with RTP, the Real-Time Streaming Protocol (RTSP) can be used. RTSP is a complementary protocol for RTP, offering extended control over the transmission of real-time media.

RTP can be used for both unicasting and multicasting applications. See Section 10.6 for more information.

10.6 Unicast, Broadcast, and Multicast

There are three different methods for transmitting data on a computer network, each catering to different needs:

1. *Unicast.* Unicast (Figure 10.12) is the most common form of communication, whereby the sender and the recipient communicate on a point-to-point basis. Data packets are sent only to one recipient and no other computers on the network will receive the information.

2. *Multicast.* Multicast (Figure 10.12) is the communication between a single sender and multiple receivers on a network. Multicast technologies are used when many receivers want the same information — for example, live video — simultaneously. Multicasting reduces network traffic by delivering a single stream of information to many recipients. The biggest difference, when compared with unicasting, is that the video stream only needs to be sent once, whereas with unicast, a copy for each recipient is required. Multicasting typically is used when large numbers of users wish to view the live surveillance video. The RTP is the most common protocol used for multicasting video streams.

3. *Broadcast.* Broadcast means that the sender is sending information to all other servers on the network. When a broadcast message is sent, all hosts on the network receive the message and process it to some extent. Too many broadcast messages will slow a network and the hosts connected to it. Routers block broadcast messages and are used to create broadcast domains that limit broadcasts to the network segment on which the broadcast originated. In broadcast addressing, a distinction is made between a limited broadcast (address: 255.255.255.255), which is not forwarded via routers, and a direct broadcast (146.15.255.255, for instance), which is forwarded, if necessary, via routers to reach all hosts. Broadcasts are not practical for network video transmissions. Network video products use broadcasts only for specific protocols that require it, such as DHCP.

10.7 Quality of Service

At present, fundamentally different applications — for example, telephone, surveillance video, and e-mail — are using the same IP network. In these networks, there is a need to control how network resources are shared to fulfill the requirements of each service. One solution is to let network routers and switches operate differently on different kinds of services (voice, data, and video) as traffic passes through the network. Using Quality of Service (QoS), different network applications can coexist on the same network without consuming each other's bandwidth.

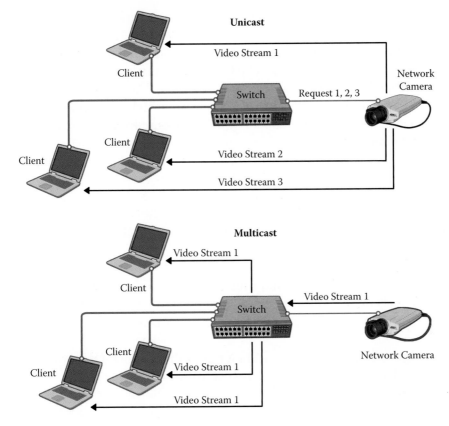

Figure 10.12 Unicast and multicast video transmission.

10.7.1 Definition

The term "Quality of Service" (QoS) refers to a number of technologies that guarantee a certain quality to different services on a network. The quality can be, for example, a maintained level of bandwidth, low latency, or no packet losses. The main benefits of a QoS-aware network can be summarized as:

- The ability to prioritize traffic and so that critical flows can be served before flows with less priority
- Greater reliability in a network by controlling the amount of bandwidth an application can use, and thus controlling bandwidth competition between applications

QoS relates to the transmission delay (latency) between systems, jitter (variation from the average latency time), packet loss rate (loss probability for individual packets), and data throughput (bandwidth).

Datagram headers in IPv4 and IPv6 contain a Differentiated Services Code Point (DSCP) flag for identifying the type of data in the relevant IP datagram. Using this flag, the data packets are divided into traffic classes and prioritized for forwarding. The DSCP flag has a length of 6 bits, which means that 64 different classes can be defined. The prerequisite for the use of QoS within a video network is that all switches, routers, and network video products must support QoS. An example of the use of QoS is shown below.

10.7.2 QoS in Network Video

To use QoS in a network with network video products, the following requirements must be met:

- All network switches and routers must include support for QoS. This is important to achieve end-to-end QoS functionality.
- The network video products must be QoS enabled.

See Figure 10.13 and Figure 10.14 for an ordinary (non-QoS-aware) network and a QoS-aware network, respectively.

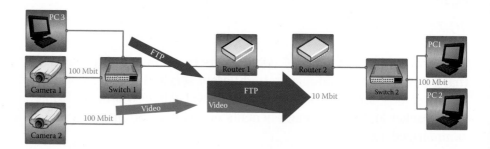

Figure 10.13 Ordinary (non-QoS-aware) network. In this example, PC1 is watching two video streams from cameras Camera 1 and Camera 2, with each camera streaming at 2.5 Mbps. Suddenly, PC2 starts a file transfer from PC3. In this scenario, the file transfer will try to use the full 10-Mbps capacity between routers 1 and 2, while the video streams will try to maintain their total of 5 Mbps. The amount of bandwidth given to the surveillance system can no longer be guaranteed, and the video frame rate will probably be reduced. At worst, the FTP traffic will consume all the available bandwidth.

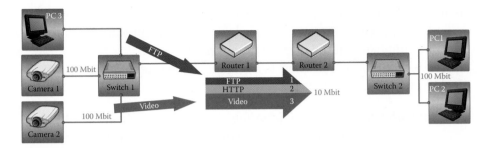

Figure 10.14 QoS-aware network. Here, Router 1 is configured to devote up to 5 Mbps of the available 10 Mbps for streaming video. FTP traffic is allowed to use 2 Mbps, and HTTP and all other traffic can use a maximum of 3 Mbps. Using this division, video streams will always have the necessary bandwidth available. File transfers are considered less important and get less bandwidth, but there still will be bandwidth available for Web browsing and other traffic. Note that these maxima only apply when there is congestion on the network. If there is unused bandwidth available, this can be used by any type of traffic. PTZ traffic is often regarded as critical and requires low latency to guarantee fast responses to movement requests. This is a typical case where QoS can be used to provide the necessary guarantees.

10.8 Network Security

Considering the wide use of IP networks for data, video, and voice by governments, banks, and enterprises, securing the transfer of information is necessary. Although early IP networks had some security flaws, a tremendous amount of R & D money was spent during the past 10 years to improve security measures, and this has resulted in extremely secure networks. There are several ways to provide security within a network and between different networks and clients. Everything from the data that is sent over a network to the actual use and accessibility of the network can be controlled and secured.

Providing secure data transmission can be likened to using a courier to deliver a valuable and sensitive document from one person to another. When the courier arrives at the sender's place, the courier would be asked to prove his identity. Once this is done, the sender would decide if he is who he claims to be and if he can be trusted. If everything seems correct, the locked and sealed briefcase would be handed to the courier, and he would deliver it to the recipient. At the receiving end, the same identification procedure would take place, and the seal would be verified as "unbroken." Once the courier has left, the recipient would unlock the briefcase and take out the document to read it.

A secure communication is created in the same way and can be divided into two different types:

1. *Authentication and Authorization.* This initial step involves the user or device identifying itself to the network and the remote end. This is done by providing some kind of identity — through a username and password, for example — to the network or system.

   The next step is to have this authentication authorized and accepted, that is, verifying whether the device has the authority to operate as requested. Once authorization is completed, the device is fully connected and operational in the system.

2. *Privacy.* Privacy is accomplished by encrypting the communication to prevent others from using or reading the data. The use of encryption can slow communications, depending on the kind of implementation and encryption used. Privacy can be achieved in several ways. Two of the more commonly used methods are VPN (virtual private network) and SSL/TLS (also known as HTTPS), and WEP (Wired Equivalent Privacy) or WPA (Wi-Fi Protected Access) in wireless networks.

The main technologies for securing data transmissions are explained below.

10.8.1 Username and Password Authentication

The most basic method of protecting data on an IP network is to use username and password authentication. Data is protected from access until a user submits the correct username and password. The passwords can be encrypted or unencrypted when they are sent; the former provides the best security.

Username and password protection may be appropriate in an installation where high levels of security are not required or where the video network is segmented from the main network and unauthorized users would not have physical access to the video network.

10.8.2 IP Filtering

Some network cameras and video encoders include IP filtering, which prevents all but one or a few IP addresses from accessing the network

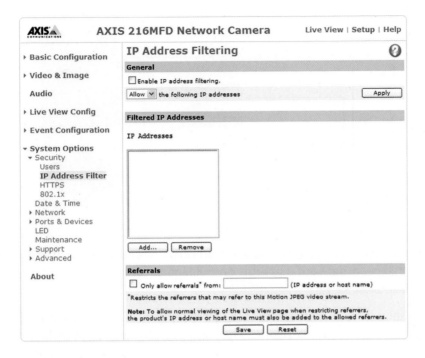

Figure 10.15 IP filtering in a network camera restricts access to only one or a few IP addresses.

video products (Figure 10.15). It provides a function similar to a built-in firewall.

Some installations that require a higher level of security can use IP address filtering. A typical configuration is to configure IP cameras to allow only the IP address of the server that is host to the video management software.

10.8.3 802.1X

Pushed by the wireless community's search for stronger security methods, the IEEE 802.1X standard was introduced and is among the most popular authentication methods in use today. It provides authentication to devices attached to a LAN port, establishing a point-to-point connection or preventing access from that port if authentication fails. 802.1X is often referred to as Port Based Network Access Control and prevents what is called "port hijacking" — that is, when an unauthorized computer obtains access to a network by getting to a network jack inside or outside a building.

Authentication is based on three instances: (1) the supplicant, (2) the authenticator, and (3) an authenticating server. The supplicant

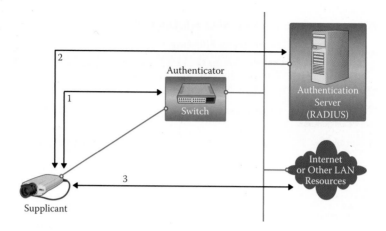

Figure 10.16 IEEE 802.1X enables port-based security and involves a supplicant (e.g., a network camera), an authenticator (e.g., a switch), and an authentication server.

corresponds to a network device, for example, a network camera that requests access to a network. The authenticator can be a switch or an access point. Logical ports on the authenticator allow user data from the supplicant to pass through once the supplicant is authenticated. The authenticating server is a (dedicated) server on the LAN to which servers must identify themselves in the authentication process.

The authenticating server is called a Remote Authentication Dial-In User Service (RADIUS), which can be a Microsoft Internet Authentication Service server. If a device wants to access a network, it asks for access to the network through an authenticator, which forwards an authentication query to an authentication server. If authentication is successful, the server instructs the authenticator to authorize access to the network for the querying server (Figure 10.16).

802.1X frequently is built into network cameras and video encoders and is very useful in network video applications. The reason is that network cameras are often located in public spaces (e.g., receptions, hallways, meeting rooms, or even outside a building). Without 802.1X, having a network jack that is openly accessible poses a substantial security risk. Additionally, in today's enterprise networks, 802.1X is becoming a basic requirement for anything connected to the network.

10.8.4 Virtual Private Network (VPN)

A virtual private network (VPN) uses an encryption protocol to provide a secure tunnel, from one network to another, through which data can

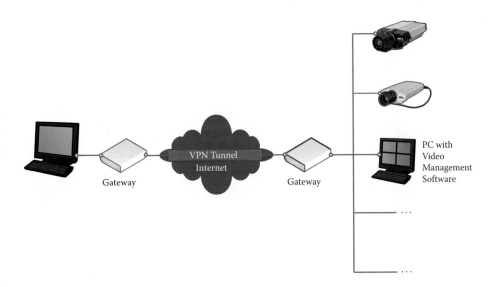

Figure 10.17 VPN security on a video network.

be securely transferred. This allows for secure communications across a public network, such as the Internet. Only devices with the correct "key" will be able to work within the VPN.

A VPN typically encrypts the packets on the IP or TCP/UDP layers and above. The IP Security (IPSec) Protocol is the most commonly used VPN encryption protocol. IPSec can use different encryption algorithms. Today, either the Triple Data Encryption Standard (3DES) or the Advanced Encryption Standard (AES) is used. AES offers higher security and needs considerably less computing power than 3DES to encrypt and decrypt data. AES uses either 128-bit or 256-bit key lengths.

Figure 10.17 illustrates VPN security on a video network.

10.8.5 HyperText Transfer Protocol Secure (HTTPS)

Another way to provide privacy is by applying encryption on a higher level — that is, the data but not the transport protocol is encrypted. HyperText Transfer Protocol Secure (HTTPS) is the most common data encryption protocol. HTTPS is commonly used in online banking to provide the requisite security for banking transactions performed over the Internet. HTTPS is identical to HTTP, but with one key difference: the data transferred is encrypted using Secure Sockets Layer (SSL) or Transport Layer Security (TLS).

Figure 10.18 shows the difference between VPN and SSL.

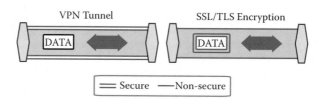

Figure 10.18 The difference between VPN and SSL is that in a VPN, the tunnel is encrypted, whereas in SSL, the actual data is encrypted. Both technologies can be used in parallel, but this is not recommended because each technology will add overhead and decrease the performance of the system.

Netscape developed and published SSL in 1994. The security offered by SSL/TLS is based on three main elements: (1) authentication of the communication partner, (2) symmetrical data encryption, and (3) protection against the manipulation of transferred data.

When making an SSL/TLS connection, negotiation first takes place through the use of a handshake protocol that determines which cryptographic methods will be used and the secure identification or authentication of the communication partner. The last element is achieved by the Web server identifying itself to the Web browser using a certificate (explained below). Then what is referred to as a premaster secret is exchanged between the communication partners over an asymmetrical encryption or Diffie-Hellman key exchange.

A certificate can be compared to an ID card that a person uses to prove his or her identity. It is a binary document usually issued by a Certificate Authority (CA), such as VeriSign. Anyone can issue certificates normally used inside closed user groups, such as a LAN Web server to which only company employees have access.

Many network video products have built-in support for HTTPS that makes it possible to view video securely using a Web browser. However, the use of HTTPS can slow the communication link and therefore the frame rate of the video.

Chapter 11

Servers and Storage

Servers and storage are important components in a network video system, as they are used for monitoring, recording, managing, and archiving video. A major benefit with IP (Internet Protocol)-based video surveillance is the ability to deploy standard server and storage components from the IT world. The components can range from a $500 desktop PC to a blade server system attached to a storage area network (SAN) with petabytes of storage that can cost millions of dollars.

Another benefit of using standard components is that they can be serviced and maintained along with other IT equipment within an organization. This not only reduces costs but also increases the availability and reliability of the equipment.

Because all the equipment consists of standard IT components, this chapter provides just a brief overview of server and storage technologies, along with some best practices relating to network video.

11.1 Servers

Depending on the performance required, there are some considerations to take into account when selecting the appropriate server platform. Because of the fast development pace within the IT industry, the rule of thumb is that the performance at a certain price point doubles every 18 months (a.k.a. Moore's law). The benefit for a network video system is that servers are easily upgraded; for example, if greater performance and capabilities are needed, the video management software can be installed on a new server.

11.1.1 Hardware Platforms

Most hardware platforms are based on Intel or AMD (Advanced Micro Devices, Inc.) processors. There are, however, some differences between a desktop platform used for word processing and e-mail and the enterprise-class servers normally used in network video applications. Enterprise servers are designed for handling large amounts of data and for continuous reading and writing. They, therefore, differ from a desktop PC in the following areas:

- *Processor type.* Compared with single-processor-based PCs, enterprise servers usually come equipped with dual- or quad-core processors or multiple processors.
- *Internal memory.* The internal RAM (random access memory) is the memory used for processing and buffering. Because a server in a video system must handle large amounts of data, large RAM buffers are necessary. An enterprise server usually has 2 GB (gigabytes) or more.
- *Storage.* Compared with a standard PC designed for occasional reading and writing, enterprise solutions require reliable storage solutions with high throughput and continuous reading and writing to serve multiple clients.

There are also other parameters within a server to consider. If the server also will be used for video monitoring, it should have an appropriate graphics card with hardware-accelerated video decoding. Servers typically are available in different footprints such as a stand-alone tower, a rack-mounted server, or a blade server. Blade servers are becoming popular because of their high density and suitability for larger video management systems.

11.1.2 Operating Systems

To make the server hardware platform operational requires an operating system (OS). In theory, any operating system can be used as a platform for recording and managing video as long as the video management software installed on top of the operating system supports it. Selecting an operating system depends on a number of technical requirements. Most organizations and companies standardize on one OS platform for their business applications to streamline administration and management.

This may dictate the video management platform. The most commonly used platforms are Windows and Linux:

- *Windows.* Windows (with its Server 2003, XP, and Vista) is the most common platform for video management applications. With Active Directory, it is possible to have centralized authentication and authorization services for Windows-based computers. The file system used in Windows is NTFS (New Technology File System), which supports a file system up to 256 TB (terabytes) and file sizes of 16 TB. NTFS has better reliability, disk utilization, and performance than its predecessor, FAT32 (File Allocation Tables).

- *Linux.* Linux is a popular UNIX-like operating system that comes in a variety of distributions. It is not commonly used in network video applications but is very popular as an embedded platform for network video devices such as network cameras and video encoders. The most common file system used is ext3fs (third extended file system), which supports file systems up to 32 TB and file sizes up to 2 TB.

11.1.3 Video File Systems

As discussed above, different file systems have different capabilities. When storing recorded video on a hard disk, it is important to have a well-organized structure for video, regardless of the operating system or file systems used. The video can be stored as standard files (JPEG, MPEG, AVI, ASF, etc.) or as raw indexed data. The structure can be a simple directory tree (with sub-directories containing files for every camera, week, day, and hour), or it can be a complex database with or without multiple references, together with metadata information.

There is no standardized way to organize large amounts of recordings. Video management software vendors use different technologies; some use standard databases such as SQL (Structured Query Language) or Oracle, or file systems, and others use proprietary formats and structures. Even two versions of the same application can differ in terms of how video is stored. The structure can be optimized for high recording performance, quick search, tampering detection, stability, and recovery, among others. Unless an application has an open interface for exporting, accessing, or searching stored video, it can be extremely difficult for other applications that use a different system to read and interpret the recordings.

11.2 Hard Disks

Single hard disks used in PC servers and desktops today range from 80 GB up to 750 GB, and drives with 1 TB (i.e., 1,000 GB) will soon be available. In addition to size, there are several other factors that differentiate hard disks. One is the spin speed, measured in rotations per minute (RPMs), which normally ranges from 5,400 to 10,000 RPMs. Another is the non-sequential (random) read/write performance of the disk system, which indicates the speed of reading and writing to non-sequential blocks of data.

Hard disks come in different sizes. For servers, a 3.5-inch standard casing is the norm. One major differentiator is the interface. The three most common interfaces — SCSI, ATA, and SATA — are described next.

11.2.1 The SCSI Interface

The SCSI (Small Computer System Interface) is a standard for physically connecting computers and peripheral devices such as hard disks and tape drives, CD drives, and DVD drives. Up to seven (or fifteen) devices can connect to one controller. The transfer rate varies between various SCSIs and can be as high as 640 Mbps (Ultra-640 SCSI). SCSI hard drives are designed and optimized for server applications that require continuous 24/7 reading and writing. The MTBF (mean time between failure) for SCSI drives is normally 1 million to 1.5 million hours.

In demanding video surveillance systems, SCSI disks are recommended for their durability as well as for their performance. Table 11.1 provides an overview of the different SCSI standards.

11.2.2 The ATA and SATA Interfaces

ATA (Advanced Technology Attachment), also known as IDE, ATAPI, and PATA (Parallel ATA), is another standard for physically connecting computers and peripheral devices such as hard disks and tape drives, CD drives, and DVD drives. It is the most common interface used in desktop and laptop environments today. Up to two devices can connect to each controller (master and slave). The transfer rate can be as high as 80 Mbps, but 66 Mbps is more realistic. The drives are designed for workstations and laptops for sporadic reading and writing. The MTBF is normally 150,000 to 800,000 hours.

Table 11.1 The SCSI Standard Has Improved Continuously, with Ever-Increasing Performance as a Result

Interface	Connector	Number of Bits (width)	Throughput (Mbps)	Number of Devices
SCSI-1	IDC50; Centronics C50	8	5	8
Fast SCSI	IDC50; Centronics C50	8	10	8
Fast-Wide SCSI	2 × 50-pin (SCSI-2); 1 × 68-pin (SCSI-3)	16	20	16
Ultra SCSI	IDC50	8	20	8
Ultra Wide SCSI	68-pin	16	40	16
Ultra2 SCSI	50-pin	8	40	8
Ultra2 Wide SCSI	68-pin; 80-pin (SCA/SCA-2)	16	80	16
Ultra3 SCSI	68-pin; 80-pin (SCA/SCA-2)	16	160	16
Ultra-320 SCSI	68-pin; 80-pin (SCA/SCA-2)	16	320	16
Ultra-640 SCSI	68-pin; 80-pin	16	640	16

SATA was designed as a successor to ATA and serves the same purpose of physically connecting computers and peripheral devices. SATA is quickly replacing ATA as the primary architecture in the PC and laptop market. Although a throughput of 300 MBps (SATA-300) is lower than SCSI, the upcoming SATA-600 will have a performance similar to the highest-performing SCSI. The MTBF for SATA drives is similar to ATA, that is, 150,000 to 800,000 hours. This does not limit them from being used in mission-critical installations as long as they are in a RAID (Redundant Array of Independent Disks) configuration (see Section 11.4.1).

SAS (Serial Attached SCSI) is a new-generation, serial communication protocol that allows for much higher speed data transfers and is compatible with SATA. SAS uses serial communication instead of the parallel method found in traditional SCSI devices. However, it still uses SCSI commands for interacting with SAS.

11.2.3 Hard Disk Failure Rates

Although failure rates for today's hard drives are much lower than for previous-generation products, according to research conducted (including reviewing hard disk drive vendors' MTBFs and third-party hard disk drive reliability studies), the most common failure mechanisms in the current generation of hard drives appear related to the head-disk interface. These problems can have many different sources, including handling damage, temporary interface disruption, media damage, and thermo-mechanical stability of the read and write structures. Regarding the timing associated with an actual hard drive failure, research suggests a failure would likely occur either in the first 60 days after the system is put into service or in the third, fourth, or fifth year of operation, given the high duty cycle and operational characteristics expected in a video surveillance storage application.

Because the loss of data can be very costly, using RAID configurations for redundancy in video management system is recommended. Doing a "burn-in" test for a few days on a new server in a video surveillance system also is recommended to ensure the drives do not fail within the first few days.

11.3 Storage Architectures

The appetite and requirement for more and more storage for all kinds of applications are driving the storage industry to develop ever-increasing performance and size of storage systems. High-quality surveillance video puts very high demands on a storage system, and storage is often a considerable part of the cost of a network video system. Virtually any size of storage system can be accommodated, which means that any frame rate, number of cameras, and retention time can be handled.

There are a few basic architectures for storage systems, all with different complexity, performance, cost, redundancy, and scalability. The most common ones are described next.

11.3.1 On-Board Storage

In some applications, storing the video on board — that is, in the memory of a network camera — can be appropriate. For example, a remote office with one or a few cameras may choose to record video locally during the day and then transfer the video over a WAN (wide area network) or the Internet at night to video management servers located in a central location. Another scenario for a distributed storage model would be a mission-critical one where the loss of video is not acceptable when the network goes down or when it is taken down for scheduled maintenance. This scenario is more hypothetical in nature because correctly designed networks today have a very high degree of reliability.

All network cameras include some level of RAM that can be used to store a few minutes' worth of video. Because of the high cost of RAM, the memory is normally quite limited. It is also volatile, which means it will be erased when the power is off. More and more cameras provide a slot for a standard memory card like an SD card (secure digital card) or enable insertion of a USB (Universal Serial Bus) thumb drive, which can add hundreds of megabytes for many hours' or even days' worth of video storage, depending on the frame rate and resolution.

11.3.2 Single Server Storage

This is probably the most common solution for hard disk storage in small- to medium-sized installations of up to 50 cameras (Figure 11.1). The hard disk is located in the same PC that runs the video management software (application server). The PC and the number of hard disks it can hold determine space availability. Most PCs with an ATA interface can hold two disks; some with two ATA buses can hold up to four disks. If using SCSI, the number of disks can theoretically be seven or fifteen, although in reality there is no space for that many. Each disk can be up to approximately 750 GB. This gives a total hard disk capacity of up to 3 TB using four disks.

In applications where the amount of stored data and management requirements exceed the limitations of direct attached storage, a separate storage system is implemented. These systems are network attached storage (NAS) and storage area network (SAN).

11.3.3 Network Attached Storage (NAS)

NAS (network attached storage) provides a single storage device directly attached to a LAN (local area network) and offers shared storage to all

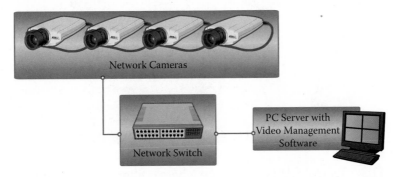

Figure 11.1 Example of a single server storage setup.

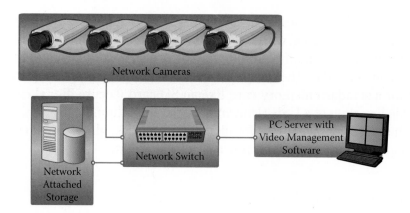

Figure 11.2   Example of a network attached storage setup.

clients on the network. A NAS device (Figure 11.2) is simple to install and easy to administer. It provides a low-cost solution for storage requirements but is limited in its data throughput.

11.3.4 Storage Area Network (SAN)

A SAN (storage area network) is a high-speed, special-purpose network for storage devices. It is connected to one or more servers via a fiber channel (Figure 11.3). Users can access any of the storage devices on the SAN through the servers, and storage is scalable to hundreds of terabytes or

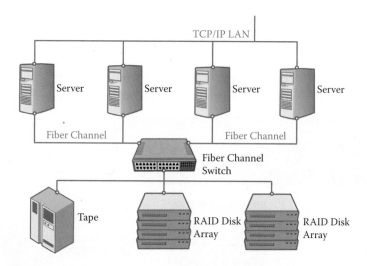

Figure 11.3   Typical SAN architecture where a fiber channel network ties all storage devices together and lets the servers share the storage capacity.

even petabytes (1,000 terabytes). Centralized data storage reduces the administration required and provides a flexible, high-performance storage pool for use in multi-server environments.

The difference between NAS and SAN is that NAS is a storage device wherein the entire file is stored on one single hard disk, whereas a SAN consists of a number of devices where the file can be stored, block by block, on multiple hard disks. This type of hard disk configuration allows for a very large and scalable hard disk solution where vast amounts of data can be stored with a high level of redundancy.

11.3.5 Internet Small Computer System Interface (iSCSI)

iSCSI (Internet SCSI) is a network protocol that makes it possible to send SCSI commands over a TCP/IP network instead of an SCSI cable. This makes it possible to connect a storage device and address it as a local drive on the server. Unlike NAS (network attached storage), iSCSI devices do not support file system protocols. The main difference between iSCSI and NAS is that iSCSI provides block-level disk access, whereas NAS only provides file-level access. The host server manages the file system, just like a local attached drive. iSCSI is commonly used in large installations, usually in RAID configurations, as it is a cost-efficient way of quickly adding more storage to a server.

11.4 Redundancy

The server and storage components are essential in any IT system, including an IP-based video surveillance system. Many different technologies are available to make the system more reliable by increasing the redundancy. The following subsections discuss the most common techniques.

11.4.1 RAID Systems

RAID (Redundant Array of Independent Drives, previously called Redundant Array of Inexpensive Drives) is a method of arranging standard, off-the-shelf hard drives in such a way that the operating system sees them as one large logical hard disk. The benefits are both increased throughput and increased reliability. Using hardware RAID controllers instead of software controllers is recommended to reduce performance issues.

There are different levels of RAID that offer different levels of redundancy — from practically no redundancy at all to a full "hot swappable" mirrored solution where there is no disruption to the operation of the system and no loss of data in the event of a hard disk failure.

The most common RAID levels include:

- *RAID-0: also called striping.* Information is spread out over two or more disks to increase the performance. There is no redundancy because one disk failure will destroy the array.
- *RAID-1: also known as disk mirroring.* The information on one disk is duplicated onto one or more disks. This increases the reliability but may reduce the performance, as data needs to be written on two disks.
- *RAID-5: also known as striping with parity.* The data and parity are spread over three or more disks and require at least three disks in the array. Read performance is the same as for a single disk. Write performance can be lower, as data needs to be written on two disks. RAID 5 can tolerate a single disk failure and still recover all data. Additionally, the disks can be made hot swappable. RAID-5 has become popular because it provides redundancy and maximizes disk space used for data instead of backup.
- *RAID-6: similar to RAID 5 but with dual parity bits.* This requires at least four disks, and the configuration can tolerate two disk failures.

11.4.2 Data Replication

Data replication is a common feature of many network operating systems; file servers in the network are configured to replicate data in each other's server (Figure 11.4).

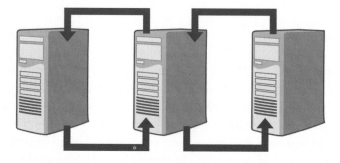

Figure 11.4 By replicating data, a high level of redundancy is achieved.

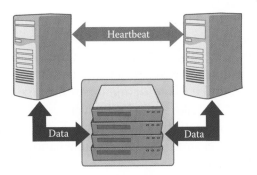

Figure 11.5 Server clustering is a popular way to ensure redundancy on the server piece of the system.

11.4.3 Tape Backup

Tape backup is an alternative or complementary method. There is a variety of software and hardware equipment available on the market, and backup policies normally include taking tapes off site as prevention against fire or theft.

11.4.4 Server Clustering

Many server clustering methods exist. A common method for database servers and mail servers is when two servers are working with the same storage device, commonly a RAID system. In such a case, when one server fails, the other one (which is configured identically) takes over the application (Figure 11.5). These servers usually share even the same IP address, making the so-called failover completely transparent for the user.

11.4.5 Multiple Servers

A common method to ensure disaster recovery and off-site storage of network video is to simultaneously send the video to two different servers located in separate locations. These servers, in turn, can be equipped with RAID, work in clusters, or replicate their data with servers even further away.

11.5 Best Practices

Selecting the right server and storage platforms for a video surveillance system requires many considerations. Some of the questions that need addressing include:

- *Has the organization standardized?* In most organizations, the type and brand of server and operating system are set, so the video management system should also run on the same type of server. The obvious benefit is that service and maintenance agreements are in place, and the IT department can manage the video management server just like any other server.

- *Is a centralized or a decentralized system preferred?* This depends on the size of the system and the available bandwidth between the different locations.

- *What is the required system reliability?* If it is high, one should use a high-quality server with SCSI drives.

- *What is the required system redundancy?* A RAID-based recording system might be a smart investment that ensures no video is lost if a hard disk fails.

- *What are the size and scalability?* How many cameras will be managed, and how scalable must the system be? Always plan for growth. The server should not be used at maximum performance and 90 percent of the storage space should not be used from day one. There will always be a need to add to the system down the road, so plan for such a system today.

- *What is the system performance?* Even if monitoring and recording at 30 fps at the highest resolution is desirable, such requirements normally make the system unnecessarily expensive. In most cases, it is difficult to see a difference between 15 fps and 30 fps, and many times 4 fps is more than enough. Use motion-based recording at night and reduce the resolution, if possible.

- *How long will the video be retained?* Going from 15 to 30 days will double the storage cost. Remember that more than 99 percent of all video is never watched. Having different retention times for different cameras is recommended, that is, a few days' retention for some cameras and longer retention times for other, more important camera locations.

- *Is a separate monitoring station needed?* Having the monitoring server separated from the recording server is normally recommended in systems with more than 30 cameras.

Chapter **12**

Video Management

Video management plays an important role in determining the cost-effectiveness and success of a video surveillance system. Without effective means of managing video, recording and proper viewing would be difficult, and too much unimportant video would be generated, which would consume too much bandwidth and storage space. Selecting the right video management system requires much consideration with regard to the degree of scalability, flexibility, and range of functionalities required.

The basic functions of video management involve recording and viewing. If a system consists of only one or a few cameras, the built-in Web interface of the network cameras and video encoders can manage viewing and some basic recording. When the system consists of more than a few cameras, using some kind of video management system is recommended.

The two main types of video management systems are (1) the network video recorder (NVR), which is a hardware box with preinstalled software, and (2) the video management software, which is installed on a standard PC server. A Windows platform or a UNIX/Linux platform serves as the basis for these systems. Either a Web-based interface that uses a standard Web browser or one that is based on a Windows client, which requires the installation of client software, provides for viewing functionality.

Video management systems come with recording and viewing functionalities as well as functionalities for finding and configuring network video devices. They also often include event handling and security. More advanced functions include video motion detection and other video intelligence as well as integration with other systems such as access control and building management.

This chapter provides an overview of the different types of video management systems, the platforms, and the most common functionalities available, in addition to some examples of integrated systems.

12.1 Hardware Platforms

There are two different types of hardware platforms for a network video management system: (1) a PC Server platform onto which a video management software is installed, and (2) an NVR platform, which is proprietary hardware with preinstalled video management software. Each platform has its advantages and disadvantages, as outlined below.

12.1.1 PC Server Platform

A PC Server platform solution is commercial off-the-shelf (COTS) hardware where hardware components, such as dual-processor systems and storage, can be selected to obtain the maximum performance for the specific design of the system. With a PC Server platform solution, it is possible to use standard components for increased or external storage, for additional remote operator stations, and for running additional software, such as firewalls and virus protection, in parallel with the video application. The video management software can be installed on a PC server by an end user or a systems integrator.

Systems designed on a PC Server platform are fully scalable. Normally, the video management software is designed so that one license per camera is required, and licenses can be added one by one. The system hardware can be expanded or upgraded to meet increased performance requirements. This platform is suitable for system scenarios where a large number of cameras are deployed or when the IT department has standard specifications on the type of server hardware and software allowed on the network. In demanding environments, the video management software can be installed on a ruggedized, industrial PC server.

12.1.2 NVR Platform

The most obvious difference between an NVR platform and a PC Server–based solution is that an NVR comes as a hardware box with preinstalled video management functionality (Figure 12.1). In this sense, an NVR is similar to a DVR (digital video recorder). (Some DVRs, often called

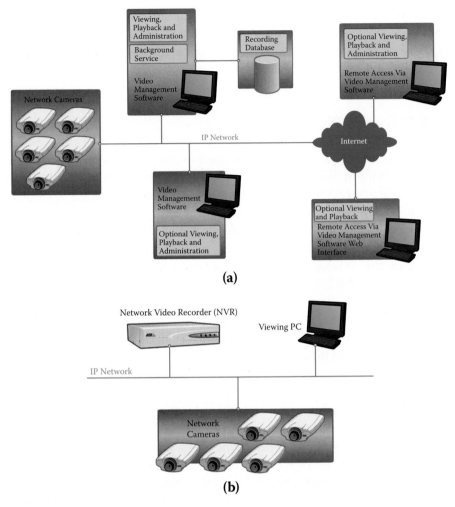

Figure 12.1 A network video system (a) based on a PC Server platform, and (b) that uses an NVR.

hybrid DVRs, also include an NVR function, that is, the additional ability to record network-based video.)

An NVR is dedicated to its specific tasks of recording, analyzing, and playing back network video. NVRs do not allow for any other applications to reside on them. The NVR hardware itself is "locked" to its application, and the unit can very rarely be altered to accommodate anything outside its original specification.

Often, the NVR hardware is proprietary and specifically designed for video management. The operating system can be Windows, UNIX/Linux, or proprietary. An NVR is designed to offer optimal performance for up to a set number of cameras and is normally less scalable than a PC

Server platform system. This makes the unit suitable for smaller system configurations where the number of cameras stays within the limits of an NVR's designed capacity. An NVR is normally easier to install in comparison with a PC Server platform.

12.2 Software Platforms

Different software platforms can be used to manage video. They include using the built-in Web interface, which exists in many network video products, or using a video management system with either a Windows-based or a Web-based interface.

Video management systems are available from vendors of network cameras and video encoders. They often support only the network video devices of the vendor. Open systems that support multiple brands of network video devices also exist, often from independent companies. Hundreds of such open video management software systems have been made available over the past few years. Some software is more scalable and includes more advanced functionalities than others. Additionally, the licensing model can differ.

12.2.1 Built-in Functionality

Most network cameras and video encoders have a built-in Web server with an IP address, so viewing the video and configuring all features in a network video product can be done by simply typing the product's IP address in the Address/Location field of a Web browser on a computer. After making a connection with the network video product, the product's "start page" automatically displays in the Web browser. The start page displays live video along with links to the product's configuration pages.

Recording of video or image snapshots is often available by right-clicking on the video in the Web browser. Configuring and managing a network video product through its built-in functionalities work when only a small number of cameras or video encoders are involved in a system (Figure 12.2).

12.2.2 Windows Client-Based Software

The most common video management systems are Windows based. For large systems, or if the viewing station is located in a remote location,

Figure 12.2 Live viewing using a Web browser and a network camera's built-in functionalities.

Windows-based viewing client software is installed on a server that is separate from the recording server where the management software is installed (Figure 12.3). The client application allows a user to perform the same tasks using the same user interface as on the computer where the management software is installed. All settings are inherited and downloaded from the video management software. In some cases, the client application also enables users to switch between different servers that have video management software installed, thus making it possible to manage video at many remote sites or in a large system.

12.2.3 Web-Based Software

Web-based video management software must be installed first on a PC platform that is used for recording video and that has a built-in Web server. It then allows users on any type of networked computer, anywhere in the world, to access the software — and thereby the network video products it manages — simply with the use of a Web browser (Figure 12.4).

In the browser, users enter the IP address or host name of the PC server where the video management software is installed and, if required, enter the username and password. Although there is no need, theoretically, for the installation of the Web-based software on the remote PC, most

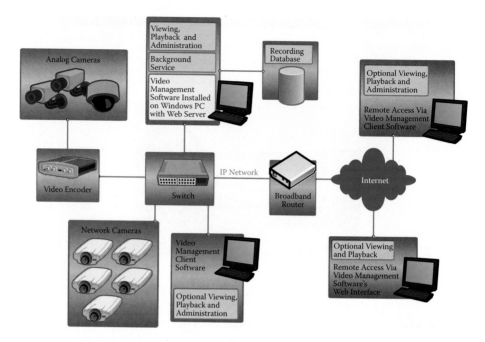

Figure 12.3 The video management software is installed on a Windows-based PC with a Web server. It allows remote users to access the video management software via a Windows client and via a Web interface.

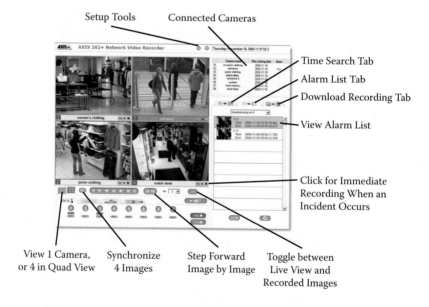

Figure 12.4 A Web-based video management software application enables users to view cameras and perform various operations with the use of a Web browser from any networked computer.

Web-based applications require the installation of Java or ActiveX components. (An ActiveX control, when used in a Windows Internet Explorer environment, is a downloadable computer program that interacts with a main application such as a Web browser to execute specific functions on demand.) Components are usually seamlessly downloaded onto the remote PC unless security measures are in place to prevent automatic downloads.

Windows client-based software often provides more functionalities and a richer interface than Web-based software. However, there are no real limitations to what a Web-based application can offer when using ActiveX controls. With the proper safeguards such as password protection and IP address filtering, Web interfaces allow for online management of video from anywhere in the world.

A video management system can be utilized for different purposes and even different departments. For example, a video surveillance system in a store can be used by one individual for security purposes, whereas another can use it for studying visitor traffic patterns.

12.2.4 Open versus Vendor-Specific Software

Virtually all network camera and video encoder vendors provide their own video management software or NVRs. Some have only very basic recording and viewing functions with limited scalability and may even be given away for free to support sales of network cameras. Others provide more full-fledged systems. Many such systems support only the network video devices of a specific vendor, which limits the choice and flexibility for the end user. A benefit, however, may be that the integration with the network cameras and encoders is more optimized because all the built-in features can be utilized.

Open systems that support network video devices from many different manufacturers often provide the highest flexibility for the end user.

12.2.5 Scalability of Video Management Software

The scalability of most video management software — in terms of the number of cameras and frames per second that can be supported — is in most cases limited by the hardware capacity rather than the software. Storing video files puts new strains on the storage hardware because it may be required to operate on a continual basis, as opposed to only during normal business hours. In addition, video, by nature, generates large amounts of data, creating high demands on the storage solution.

Recording video and audio is not very CPU (central processing unit) intensive. Much processing, however, is required if frames require decoding, as with functionalities such as video motion detection and other video intelligence that reside on the server. Live viewing capacity is limited by the capacity of the video decoder and graphics card.

As a system grows and eventually exceeds the capacity of a single server, it is in many cases possible to add new servers to the system — ideally in a seamless way. For professional and scalable systems, factors such as redundancy, failover, and no single point of failure become very critical. Redundancy in a storage system allows for saving video, or any other data, simultaneously in more than one location. This provides a backup for recovering video if a portion of the storage system becomes unreadable. There are a number of options for providing this added storage layer in a network video system, including a Redundant Array of Independent Disks (RAID), data replication, server clustering, and multiple video recipients. See Chapter 11 for additional information on servers and storage.

12.2.6 Licensing of Video Management Software

Most video management software applications are licensed products. Licensing policies vary, but in most cases they involve a license per camera or video source in the system as well as a base fee to acquire the software. The license also can be a software key tied to the MAC address of a camera or a server CPU. Additionally, there are typically maintenance fees associated with technical support and the right to download and install new versions of the software.

12.3 System Features

A video management system includes many different features. Some of the more common ones include:

- Recording of video
- Recording of audio
- Simultaneous viewing of video from multiple cameras
- Event management functions
- Camera administration and management
- Search options and playback
- User access control and activity (audit) logging
- Intelligent video applications such as video motion detection

12.3.1 Recording

One of the key functions in a video management system is recording video, as well as audio. Recording functionalities include setting up the rules of recording as well as intelligent ways of searching for recorded video and exporting video to other systems. The following subsections discuss the various recording aspects of a video management system.

12.3.1.1 Video Recording

Video can be recorded continuously, on trigger (by motion or alarm), and on schedule, which can combine both continuous and triggered recording instructions (Figure 12.5).

Continuous recording normally uses more disk space than an alarm-triggered recording. An alarm-triggered recording can be activated by, for example, video motion detection or an external input through a camera's input port. With scheduled recordings, timetables for both continuous and alarm or motion recordings can be set (Figure 12.6).

After selecting the type of recording method, the quality of the recordings can be determined by selecting the video format (e.g., Motion JPEG or MPEG-4), resolution, and level of image compression. These parameters, as well as frame rate, will affect the amount of bandwidth used as well as the size of storage space required.

The number of frames per second can be set in all recording modes. Network video products can have varying frame rate capabilities, depending on the resolution. Full-motion video is 30 frames per second in NTSC video standard (in North America and Japan) and 25 frames per second in PAL video standard (in Europe). Some network cameras have the capability to do even higher frame rates. In a well-designed system, the appro-

Figure 12.5 An example of an interface for editing recording functions.

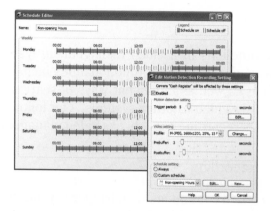

Figure 12.6 Scheduled recording settings with a combination of continuous and alarm or motion recordings applied.

priate frame rate is requested of the network camera or video encoder, which means that only the required frame rate is sent over the network.

A video management software application usually has a background service that automatically starts running upon system start-up. When the background service is running, recording will continue even after a user has logged off the PC where the video management software is installed.

12.3.1.2 Audio Recording

Audio recordings also can be made using the built-in functionality in a network video product or via video management software. When having synchronized audio and video is important, time-stamping of the audio and video packets is required. Because the time-stamping of video packets using Motion JPEG may not always be supported in a network camera, the recommended video compression to use is MPEG-4 or H.264 because such video streams are usually sent using RTP (Real-time Transport Protocol), which time-stamps the video packets. For details on audio compression standards, see Chapter 6 on audio technologies. There also can be restrictions in some markets with regard to audio recording. See Chapter 15 for legal considerations.

12.3.1.3 Recording and Storage

Most video management software uses the standard Windows file system for storage, so any system drive or network attached drive can be used for storing video. An index of available video usually is stored in a separate file or database. The advantages of using a database for storing all settings and recording metadata, and using a file system, include:

- The ability to manage shared access and ensure data integrity
- The possibility to efficiently search for recordings
- The ability to enable direct file access and record directly to disk

Video management software can enable more than one level of storage — that is, recording on a primary hard drive and archiving on local disks, network-attached drive, or remote hard drive. Users may be able to specify how long images should remain on the primary hard drive before they are automatically deleted or moved to the archive drive. Users also may be able to prevent video from being deleted automatically by locking them in the system.

12.3.1.4 Advanced Search

Searching for recorded video is an important feature in a video management system. There are different ways of finding the footage of interest. A certain date and time can be entered, and the video for one or several cameras can be replayed.

If the user knows only that a person entered a building and not the time of entry, a motion-based search can be done (Figure 12.7). By defining an area of a scene where motion is to be detected, a video management

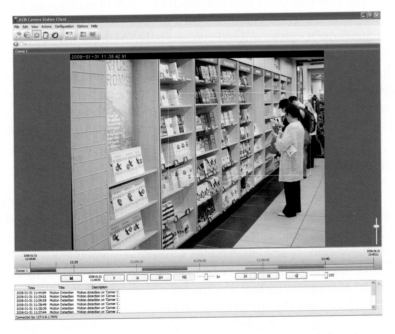

Figure 12.7 Using motion-based search, events in a defined area of a scene can be easily found by a video management system.

system can search for and replay images where there is activity in the defined area. This functionality is often called advanced search, quick search, or smart search. Video associated with a certain type of alarm or event trigger also can be easily found and replayed.

12.3.1.5 Exporting Files

Many video management software applications record video using proprietary file formats but enable users to export recorded video to standard file formats such as AVI (Audio Video Interleave) and ASF (Advanced Streaming Format) files. The proprietary format, in most cases, also can be exported along with a proprietary player. The advantages of using the native recording format are additional security and data integrity as well as advanced playback features. The proprietary format can be encrypted and password protected. Additionally, the proprietary format can be more difficult to edit and can be used to preserve the chain of evidence. Other features also may be available in a proprietary player, such as multi-camera playback.

12.3.2 Viewing

Another key function of a video management system is enabling the viewing of live and recorded video in efficient and user-friendly ways. Most video management software applications enable multiple users to view several different cameras at the same time and allow recordings to take place simultaneously. Additional features may be multi-monitor viewing and mapping, which overlays camera icons that represent the locations of cameras on a map of a building or area. The following subsections discuss the viewing functionalities commonly found in video management systems.

12.3.2.1 Live Viewing

Many video management software applications provide users with the option of viewing images in different ways: split view of from four to 64 cameras (Figure 12.8), one-camera pop-up (or pop-up on motion detection), full screen, or camera sequence mode. Camera sequencing is a predefined "tour" that automatically displays live views, one after another, from a predefined list of cameras included in the tour. In this mode, users are able to select in which order the cameras should be viewed and for how long.

When working with a PTZ (pan, tilt, zoom) or network dome camera, a video management software program may allow the PTZ function of the camera to be controlled in a number of ways:

Figure 12.8 An example of a split view.

- By clicking on a display keypad
- By using a mouse to click on the image and move the camera, and using the mouse scroll wheel to zoom in
- By using a compatible joystick

If a camera is equipped with audio capability, the audio controls also can be shown in the software interface.

12.3.2.2 Viewing of Recordings

Many video management software programs also offer a multi-camera play-back feature that enables users to view simultaneous recordings from different cameras (Figure 12.9). This provides users with an ability to obtain a comprehensive picture of an event and is helpful in an investigation.

12.3.2.3 Multi-Streaming

Viewing or recording at full frame rate on all cameras at all times is more than what is required for most applications. Frame rates under normal conditions can be set lower — for example, one to four frames per second — to dramatically decrease storage requirements. In the event of an alarm — for example, if video motion detection or an external sensor is triggered — the recording frame rate can be increased automatically. It also is possible to

Figure 12.9 By having the ability to view several cameras being replayed simultaneously, the operator may be able to get a better overview of when and how a certain incident transpired.

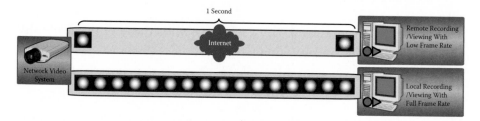

Figure 12.10 Different frame rate video can be sent to different recipients.

send video with different frame rates, compression, and resolution to different recipients (Figure 12.10). This is called multi-streaming.

12.3.2.4 Mapping Functionality

To easily select a camera, camera icons can be placed on a map that is imported into the video management software (Figure 12.11). The map file can be a photo or a drawing in a standard image format such as jpg,

Figure 12.11 Using a mapping functionality, finding the right camera becomes very intuitive. (Picture courtesy of Genetec Inc.)

gif, or png. The video management software may have an icon library that makes it possible to drag and drop icons onto a map.

The icons can represent different types of cameras. By clicking on a camera icon, live video from that camera can be displayed. When an alarm occurs, a camera icon may change color to indicate that the alarm originates from that camera. More advanced systems also make it possible to show the area covered by a camera.

12.3.3 Event Management

Video management software or a network video recorder usually has the ability to receive, process, and associate events from different systems. Events can be received from access control, point-of-sale terminals, intelligent video software, and the network video products themselves. Once an event is triggered, the video management system can register the event, associate it with a video clip from a nearby camera, and alert an operator or investigator by sending a notification to a cell phone or PDA (personal digital assistant) or by having a window pop up on a viewing terminal.

The subsections below provide more details about event and alarm management, video motion detection, input and output ports, and event log files.

12.3.3.1 Edge-Device Events

Having built-in event management functions in network video products provides tremendous benefits. It enables more efficient use of bandwidth and storage space because there is no need for a network camera or video encoder to send any video for viewing or recording unless an event takes place. In addition, live monitoring of cameras is not required all the time. When an event takes place, alerts and notifications can be sent, and all configured responses can be activated automatically.

Event management, which includes alarm handling, involves defining an event that activates a network video product to perform certain actions. An event can be scheduled or triggered. Events can be triggered by, for example:

- *Input port(s).* The input ports on a network camera or video encoder can be connected to external devices such as a motion sensor or a door switch (Figure 12.12).
- *Manual trigger.* An operator can make use of buttons to manually trigger an event.
- *Video motion detection.* When a camera detects certain movement in a camera's motion detection window, an event can be triggered.
- *Audio trigger.* This enables a camera with built-in audio support to trigger an event if it detects audio below or above a certain threshold. (For more information on audio detection, see Chapter 6.)

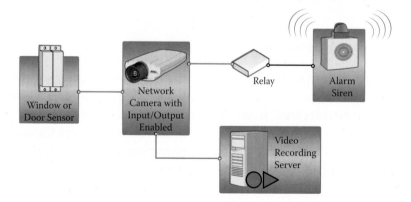

Figure 12.12 A camera is attached to a window switch via its input port and to an alarm system/siren viat its output port.

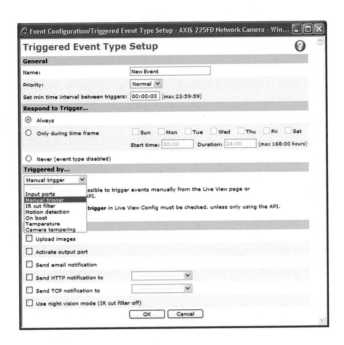

Figure 12.13 Setting event triggers.

- *Temperature.* If the temperature rises or falls outside the operating range of a camera, an event can be triggered.
- *Camera tampering.* This feature, which allows a camera to detect when it has been intentionally covered, moved, or is no longer in focus, can be used to trigger an event. (More information on camera tampering appears in Chapter 14.)

See Figure 12.13.

12.3.3.2 Responses

Network video products can be configured to respond to events all the time or at certain set times. When an event is triggered, some of the common responses that can be configured include the following:

- Upload images for recording at specified locations and at a certain frame rate using a specified compression type and level during the course of an event.
- Activate output port: the output ports on a network camera or video encoder can be connected to external devices such as alarms. (More details are provided below.)

- Send e-mail notification: this notifies users that an event has occurred. An image of the video clip also can be attached in the e-mail.
- Send HTTP or TCP notification: this is an alert to a video management system, which can then, for example, initiate recordings.
- Go to a PTZ preset: this feature may be available with PTZ or dome cameras and enables the camera to point to a specified position, such as a window, when an event takes place.
- Send an SMS with text information about the alarm or an MMS with an image showing the incident.
- Activate an audio alert on the video management system.
- Enable on-screen pop-up, showing views from a camera where an event has been activated.
- Show procedures that the operator should follow.

In addition, pre-alarm and post-alarm image buffers can be set, enabling a network video product to send a set length and frame rate of video captured before and after an event is triggered. This can be beneficial in helping provide a more complete picture of an event.

12.3.3.3 Video Motion Detection (VMD)

Video motion detection (VMD) is a common feature in video management systems. It is a way of defining activity in a scene by analyzing image data and differences in a series of images.

VMD in video management software. Video management software can provide the video motion detection functionality to network cameras that do not have this as a built-in feature. This means that a network camera will send video to the software program for analysis. Using VMD helps in prioritizing recordings, decreasing the amount of recorded video, and making searching for events easier.

With VMD, motion detection can be detected in any area of an image. In addition, users may be able to set different motion detection sensitivities for low or normal light conditions (Figure 12.14). Once motion is detected, the software can trigger an external device (such as a door to open or close, a light to turn on or off), initiate recordings from selected cameras, and send e-mail alerts. Alerts also can be triggered if motion stops, which is helpful, for example, in factory situations.

It is important to note that performing VMD using a video management software program is a CPU-intensive process and puts a heavy strain on the system if VMD is applied on many video channels.

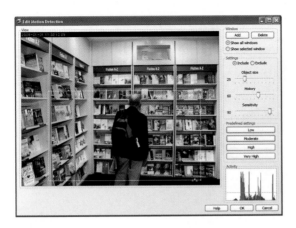

Figure 12.14 Setting video motion detection in video management software.

Embedded VMD in network cameras and video encoders. Using the built-in VMD function in a network camera or video encoder offers substantial advantages over using the VMD functionality in a software program. Because the VMD is processed in the network camera or video encoder itself, it alleviates the workload for any recording devices in the system. It also reduces the use of bandwidth, in addition to storage space, and makes event-driven surveillance possible because no video (or only low-frame-rate video) is sent to the operator or recording system unless activity is detected in a scene.

The built-in video motion detection feature in network cameras (Figure 12.15) or video encoders is very similar to the VMD functionality found in video management software. It may enable users to configure motion in certain areas while ignoring motion in others.

VMD data that provides information about, for example, the level of activity or size of a moving object also can be included in a video stream to simplify activity searches in the recorded material. More details about VMD are discussed in Chapter 14.

12.3.3.4 Input and Output Ports

A unique feature to network cameras and video encoders, in comparison with analog cameras, is their integrated input and output (I/O) ports. These ports enable a network video product to connect to external devices and enable the devices to be manageable over a network. For example, a network camera or video encoder connected to an external alarm sensor via its input port can be instructed to only send video when the sensor triggers.

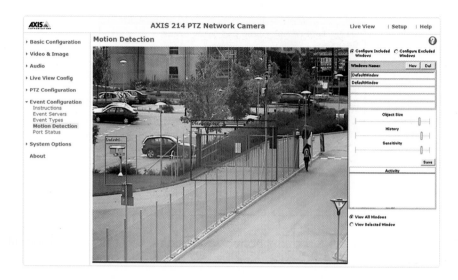

Figure 12.15 Configuring VMD in a network camera.

The range of devices that can connect to a network video product's input port is almost infinite (Table 12.1). The basic rule is that any device that can toggle between an open and closed circuit can be connected to a network camera or a video encoder. Meanwhile, the main function of a network video product's output port is to trigger external devices, either automatically or by remote control from an operator or a software application (Table 12.2).

Table 12.1 Examples of Devices That Can Be Connected to the Input Port

Device Type	Description	Usage
Door contact	Simple magnetic switch that detects the opening of doors or windows	When the circuit is broken (door is opened), the camera can take action by sending full-motion video and notifications
Passive infrared detector (PIR)	A sensor that detects motion based on heat emission	When motion is detected, the PIR breaks the circuit and the camera can take action by sending full-motion video and notifications
Glass break detector	An active sensor that measures air pressure in a room and detects sudden pressure drops (the sensor can be powered by the camera)	When an air pressure drop is detected, the detector breaks the circuit and the camera can take action by sending full-motion video and notifications

Table 12.2 Examples of Devices That Can Be Connected to the Output Port

Device Type	Description	Usage
Door relay	A relay (solenoid) that controls the opening and closing of door locks	The locking/unlocking of an entrance door can be controlled by a remote operator (over a network)
Siren	Alarm siren configured to sound when alarm is detected	The camera can activate the siren when motion is detected either using the built-in VMD or using "information" from the digital input
Alarm/intrusion system	An alarm security system that continuously monitors a normally closed or open alarm circuit	The camera can act as an integrated part of the alarm system that serves as a sensor, enhancing the alarm system with event-triggered video transfers

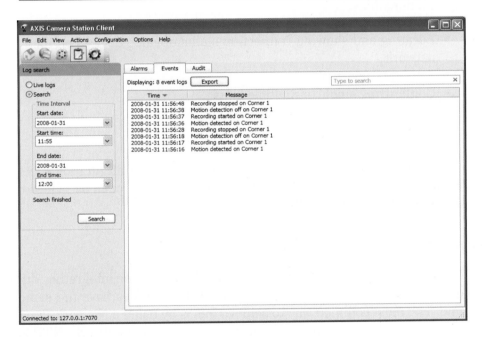

Figure 12.16 Example of an event log.

12.3.3.5 Event Log Files

Video management software can provide an event log (Figure 12.16) that provides a list of camera and server events based on date, time, type, and source of the events and that enables users to sort or search for, for example, a list of I/O activities or when motion is detected.

12.3.4 Administration and Management Features

A video management system should be able to handle the administration of cameras, such as installation, firmware upgrades, security, audit log, and parameter configurations. The larger a video surveillance system becomes, the more important it is to be able to efficiently manage networked devices.

12.3.4.1 Managing Cameras

All video management software applications provide the ability to add and configure basic camera settings, frame rate, resolution, and compression format, but some also include more advanced functionalities, such as camera discovery and complete device management.

Software programs that help simplify the management of network cameras and video encoders in an installation often provide the following functionalities:

- Locating and showing connection status of video devices on the network
- Setting IP addresses
- Configuring single or multiple units
- Managing firmware upgrades of multiple units
- Managing user access rights

Video management software also can provide a configuration sheet (Figure 12.17) that enables users to obtain, in one place, an overview of all camera and recording configurations.

12.3.4.2 Time Synchronization

Conducting an investigation using multiple cameras or integrating different systems becomes easier if all networked devices have the same time. The most common way to achieve this is by using a Network Time Protocol (NTP) server to synchronize all devices in a network. NTP is supported by most networked devices today (Figure 12.18) and can provide accurate and reliable time synchronization for a system.

12.3.4.3 Security

An important part of video management is security. A network video product or video management software should enable the following to be defined or set:

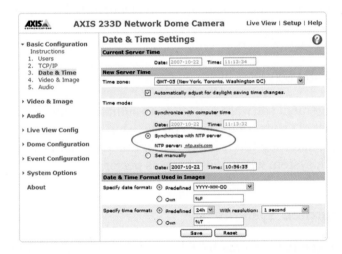

AXIS Camera Station Configuration Sheet

Generated: 15-11-2006 13:07:55
Entry Assembly: VideoMain
Application Version: 2.1.011
Application Culture: en
OS Culture: en-US
.NET Version: 1.1.4322.2032
Operating System: Win32NT
OS Version: 5.1.2600.0

Camera Settings

Name	Video Source ID	Type	DBIndex	Enabled	LAN Address	WAN Address	Video Port	Use Master Password	Video Streaming Format	PTZ Enabled	Audio Enabled
Server room	4CE0	AXIS 211	0	False	10.93.137.246	10.93.137.246	1	True	mjpg	False	False
USA	71F1	AXIS 213	1	True	72.26.147.247	72.26.147.247	1	False	mjpg	True	True
Office	C7S7	AXIS 221	2	True	10.93.10.221	10.93.10.221	1	True	mjpg	False	False
Hall	BESE	AXIS 212 PTZ	3	True	10.92.11.111	10.92.11.111	1	True	mjpg	True	False

Recording Settings

Name	Video Streaming Format	Continuous Enabled	Rec On Motion	Rec On IO	Path	Path Drive	Target Fps For Continuous	Target Fps For Alarm	Quality	Resolution	Pre Event Buffer Seconds	Post Event Buffer Seconds
Server room	mpeg-4	False	True	False	C:\Recording\	C	4	10	21	640×480	3	5
USA	mjpg	True	False	False	C:\Recording\	C	2	10	50	CIF	3	5

Figure 12.17 Example of a configuration sheet.

Figure 12.18 Most network cameras and video encoders have support for NTP.

- Authorized users
- Passwords
- Different user-access levels (Figure 12.19), for example:
  - Administrator: access to all functionalities
  - Operator: access to all functionalities except for certain configuration pages
  - Viewer: access only to live video
- Which authorized users have access to which cameras

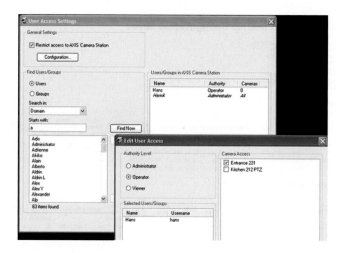

Figure 12.19 An example interface where user access can be set.

A video management software program may be able to inherit a Windows user database (local or LDAP/Domain). This is an advantageous feature because it eliminates the need to set up and maintain a separate database of users.

12.3.4.4 Audit Log Files

Video management software may offer an audit log (Figure 12.20), which presents a list of user actions based on the user, time, type of activity, and camera. The audit log file is an essential function that provides proof of who used the system when and what was performed.

12.4 Integrated Systems

Video management systems based on a network video platform can be easily integrated into other IP-based systems such as point-of-sale, access control, building management, and industrial control systems. When video is integrated, information from other systems can be used to trigger functions such as event-based recordings in the network video system, and vice versa. In addition, users can benefit from having a common interface for managing different systems.

12.4.1 Application Programming Interface

An application programming interface (API) facilitates the development of customized applications. A video management system must

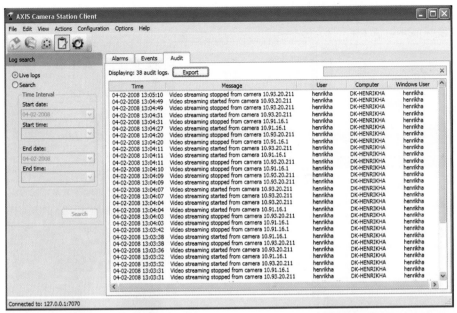

Figure 12.20 An audit log allows users to generate a list of user actions. Fields such as the user, time, type of activity, and camera can be filtered or sorted.

have an API to integrate the video system into other systems. Once implemented in another system, recordings can be triggered, alarms can be noticed, and live and recorded video from the video management system can be accessed.

12.4.2 Point of Sale

The introduction of network video in retail environments has made the integration of video with point-of-sale (PoS) systems easier.

The integration enables the linkage of all cash register transactions to actual video footage of those transactions. It helps catch and prevent fraud and theft from employees and customers. PoS exceptions such as returns, manually entered values, line corrections, transaction cancellations, co-worker purchases, discounts, specially tagged items, exchanges, and refunds can be visually verified with the captured video. It can resolve questions such as whether the right amount was entered for the products placed on the counter, whether all items on the counter were scanned, whether a return was handled properly, whether an employee discount was given to a friend, and what a person using a stolen credit card looked like. High-quality video from network cameras can provide

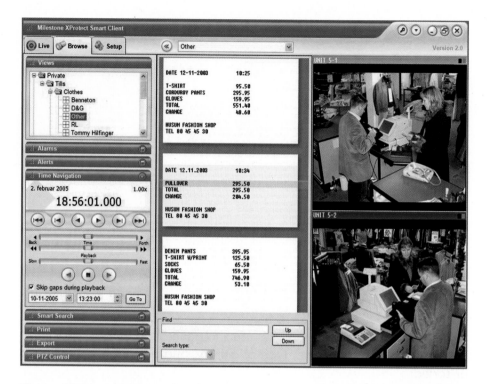

Figure 12.21 An example of a PoS system integrated with video surveillance. (Picture courtesy of Milestone Systems.)

the necessary details to identify and verify, for example, the value of a bill or items handed over to a cashier. A PoS system with integrated video surveillance makes it easier to find and verify suspicious activities.

There are several systems today that can catch PoS exceptions and store and display the receipts together with video clips of the events. Searching and viewing such events are then possible (Figure 12.21).

A PoS transaction or exception also can be used to trigger a camera to record and tag the recording. The opening of a cash register drawer, for example, can be used to trigger recordings. The scene prior to and following an event can be captured using pre- and post-event recording buffers. Such event-based recordings increase the quality of the recorded material, reduce storage requirements, and reduce the amount of time needed to search for incidents.

A PoS system with integrated video should provide the following:

- Give good insight into the various payment transactions
- Help reduce internal and external shrinkage (theft)
- Have a preventive influence on co-workers
- Show how and when mistakes are made

Figure 12.22 An example of an integrated video surveillance and access control system. (Picture courtesy of AMAG.)

12.4.3 Access Control

Integrating a video management system with a facility's access control system (Figure 12.22) allows for logging facility and room access with video. For example, video can be captured at all doors when someone enters or exits a facility. This allows for visual verification when exceptional events occur.

In addition, identification of tailgating events can be very easy. Tailgating occurs when the person who swipes his or her card, for example, knowingly or unknowingly enables others who did not swipe any card to gain entry. If a problem occurs in a secure access room, it might become important to have the ability to verify if any tailgating event occurred by looking at the recorded video and comparing it with the access control system to see if it logged everyone who entered the area. Additionally, all pictures in a badging system can be accessible to the operator of the video surveillance system for quick identification of employees or visitors.

12.4.4 Building Management

Video can be integrated into a building management system (BMS) that controls a number of systems ranging from heating, ventilation, and air

Figure 12.23 An example of a building management system that provides a single interface for monitoring a facility. The top left-hand screen provides the operator with a quad view of different camera views; the bottom left-hand screen provides an alarm summary with triggered video recordings that can be retrieved directly from an event; the top right-hand screen shows a campus graphic that enables an operator to navigate graphically throughout the system; and the bottom right-hand screen shows an air handling unit and the status and values of different controls. (Picture courtesy of Honeywell Building Solutions.)

conditioning (HVAC) to security, safety, energy, and fire alarm systems (Figure 12.23).

The following are some application examples:

- An equipment failure alarm can trigger a camera to show video to an operator, in addition to activating alarms at the BMS.
- A fire alarm system can trigger a camera to monitor exit doors and begin recording for security purposes.
- Intelligent video can be used to detect reverse flow of people into a building due to an open or unsecured door from events such as evacuations.
- Information from the video motion detection functionality of a camera that is located in a meeting room can be used with lighting

and heating systems to turn them off once the room is vacated and save energy.

12.4.5 Industrial Control Systems

Remote visual verification often is beneficial and required in complex industrial automation systems (Figure 12.24). By having access to network video using the same interface as for monitoring a process, an operator does not have to leave the control panel to visually check on part of a process. In addition, when an operation malfunctions, the network camera can be triggered to send images. In some sensitive clean-room processes, or in facilities with dangerous chemicals, video surveillance is the only way to have visual access to a process. The same goes for electrical grid systems with a substation in a very remote location.

12.4.6 Radio-Frequency Identification (RFID)

Tracking systems that involve RFID (radio-frequency identification) or similar methods are used in many applications to keep track of items. An

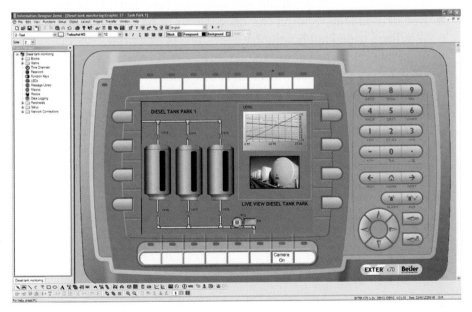

Figure 12.24 Network video integrated with an industrial control system enables an operator to visually verify an activity remotely using the same user interface. (Picture courtesy of Beijer Electronics.)

example is luggage handling at airports that will keep track of the luggage and direct it to the correct destination. If it is integrated with video surveillance, there is visual evidence when luggage is lost or damaged, and search routines can be optimized.

12.5 Best Practices

The video management system is the interface to the entire video surveillance system and is therefore a very important component. Today, several hundred different systems are available from different vendors, making choosing the right one a challenge. Selecting the right video management system and platform requires a lot of consideration, including:

- *Hardware platform.* Is an NVR platform or an open PC server platform preferred?
- *Scalability.* Some systems have limited scalability but are easy to install and operate, whereas others scale to thousands of cameras but may be complex to use in a small system.
- *Functionality.* Is a basic system that enables recording and viewing of a few cameras sufficient, or is a more advanced system with, for example, event handling and mapping functionality required?
- *Web or Windows.* Should the user interface be a Web-based or a Windows-based application? Ease of use is essential, especially if many operators are going to use it.
- *Open or vendor specific.* Most network camera and video encoder vendors supply their own video management systems that are normally limited to only one brand of products. An open system from an independent company may provide better flexibility.
- *Integration.* Is integration with a PoS, a building management, or an industrial control system important?

Chapter 13

Intelligent Video

These days, a massive amount of video is being recorded but never watched or reviewed due to the lack of time and the sheer volume of recordings. As a result, events and incidents may be missed, and suspicious behavior may not be noticed in time to prevent crime from happening.

This has led to the development of "intelligent video" (IV) applications (alternatively called video analytics, video content analysis, or intelligent video analysis). New IV systems currently are being developed to take video, convert it to data, and then analyze the data to glean interesting events. For example, automobile license plates can be recognized automatically or virtual fences can be created around critical infrastructure facilities so that alerts can be sent in the event of an intrusion.

IV systems are never idle. They can "watch" the video 24/7 in real-time, looking for events or threats, and act immediately by initiating recording or alerting security staff. IV applications can significantly reduce demands for network bandwidth and storage space, liberate staff from constant monitoring of numerous cameras, and enable rapid search to locate interesting events.

Additionally, IV systems can extract video and data from surveillance video streams and integrate the information into other applications such as retail management or access control systems, creating new benefits and opening up a wide array of business possibilities.

This chapter describes the basics and history of IV, as well as the architectures and standards applicable to IV. The following chapter, Chapter 14, discusses the most common intelligent video applications.

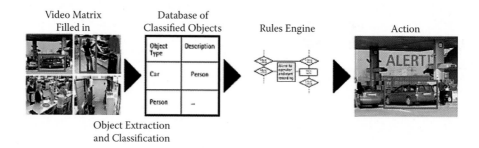

Figure 13.1 Intelligent video extracts actionable data from video images or streams.

13.1 What Is Intelligent Video?

Intelligent video or video analysis is the process of analyzing video data with the goal of transforming it into actionable information. IV systems work using complex mathematical algorithms to analyze the video and convert it into data. Typical systems extract moving objects or other recognizable forms while filtering out irrelevant movements (Figure 13.1).

The resulting data is stored in databases, which then are searched by applying a set of rules, which can be, for example, an object passing a virtual "fence" in the video or more than ten cars waiting in a drive-through line. Intelligent decision-making rules can be programmed to help determine if the events observed in the video are normal or if they should be flagged as alerts to security staff or police.

Intelligent video is quickly becoming a vital component of critical security installations, supporting timely decision making in critical situations. It also is opening up new business opportunities through applications such as people or traffic counting.

13.2 The Genesis of Intelligent Video

In 1997, the Defense Advanced Research Projects Agency (DARPA) Information Systems Office in the United States began a three-year program to develop Video Surveillance and Monitoring (VSAM) technology. The objective of the VSAM project was to develop automated video and enable it to understand and evaluate the information received for use in battlefield surveillance applications. The technologies developed under this project enable a single human operator to monitor activities over a large area using intelligent video analysis. The IV systems were designed to be autonomous, notifying the operator only if security threats occurred.

Researchers at many universities, including CMU (Carnegie Mellon University) and MIT (Massachusetts Institute of Technology), were chosen to develop a wide range of advanced surveillance techniques. They include the following:

- Real-time moving object detection and tracking from stationary and moving camera platforms
- Recognition of generic object classes (e.g., human, sedan, truck) and specific object types (e.g., campus police car, FedEx van)
- Object pose estimation with respect to a geospatial site model
- Active camera control and multi-camera cooperative tracking
- Human gait analysis
- Recognition of simple multi-agent activities
- Real-time data dissemination
- Data logging
- Dynamic scene visualization

Many of the IV companies and technologies were spin-offs from the universities involved in the VSAM project.

13.3 Why Intelligent Video?

The security industry is a rapidly growing and evolving industry. As video surveillance installations expand both in number and size, there is market demand for smarter software systems that enable management and security staff to successfully tackle their surveillance challenges. In addition, as the security market shifts from proprietary, closed analog CCTV (closed-circuit television) systems toward open, fully digital, IP-based network video, new possibilities for harvesting non-security-related information from the surveillance systems are emerging and providing new user benefits. The emergence of network video greatly simplifies the process of integrating video streams with other IT and IS applications.

The three main market drivers for intelligent video can be summarized as:

1. The limited attention span of human operators that leads to security risks
2. Speedier retrieval of stored video
3. New applications and user benefits

13.3.1 Limited Attention Span

Video surveillance in a typical security installation is limited in its effectiveness because of one major problem: it is almost impossible to watch all the video all the time. For installations with a large number of cameras, it is obviously impossible for one or even several operators to watch all the cameras. Even in the unlikely scenario that there is one security guard for every camera, a study published by Security Oz in its October/November 2002 issue found the following: "After 12 minutes of continuous video monitoring an operator will often miss up to 45 percent of screen activity. After 22 minutes of viewing, up to 95 percent is overlooked." If video is not watched and acted upon, there is obviously an increased security risk to people and facilities.

IV presents a solution to this. An IV application analyzes and filters the massive amount of information in multiple continuous video streams, ensuring that only relevant alerts are presented to security staff or police. With IV, fewer security operators can be employed to monitor even very large video surveillance installations because staff will not be expected to attentively watch dozens or even hundreds of monitors for hours on end to detect undesired activity or suspicious persons. Instead, an IV system can do the job of alerting operators when, for example, people move into restricted areas, cars drive the wrong way, crowds gather, or tampering with a video surveillance camera occurs.

13.3.2 Retrieving Stored Video

Finding incidents in stored video is extremely time consuming because the operator actually must sit and watch the recorded video. Even if the operator is experienced and can watch video at four times the speed of real-time, large archives of video can take a long time to search manually. Given this difficulty, most video files are simply stored and deleted. A major retailer's study showed that only approximately one percent of its recorded video is ever watched. In some applications, the percentage is even less.

IV systems such as video motion detection applications ensure that only relevant video footage is recorded and stored. This minimizes the need for network bandwidth and storage space and reduces the amount of relevant video data that must be searched. In addition, some IV systems can automatically search through days or even months of stored video to find security incidents in a matter of seconds.

13.3.3 New Applications

Finally, IV makes it possible to use video for applications outside security. For example, in retail stores it can be used to analyze consumer behavior. For instance, by applying a people-counting program, users can find out how many people stopped at a particular merchandising shelf. In this way, IV makes it possible to extract greater benefits from a video surveillance installation, enabling a higher return on investment.

13.4 Intelligent Video Architectures

There are two broad categories of architectures for implementing IV systems: (1) centralized and (2) distributed (Figure 13.2). In centralized architectures, video and other information are collected by cameras and sensors and brought back to a centralized server for analysis. In distributed architectures, the network cameras or video encoders, or other network components (e.g., switches), are "intelligent" and are capable of processing the video and extracting relevant information.

By intelligently designing an IV system and distributing the load, the overall costs of a system can be substantially lowered and the performance improved. The following subsections look at centralized and

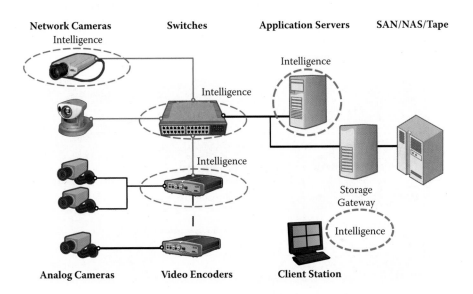

Figure 13.2 The intelligence can be located in different parts of a video surveillance system, creating a centralized or distributed architecture.

decentralized systems and how different components of a network can be used to implement IV.

13.4.1 Centralized Systems

In centralized architectures, all the video from the cameras is brought back to the "head-end" for centralized processing. Legacy infrastructures with analog cameras mostly use traditional multi-function DVRs (digital video recorders) at the head-end, whereas in a network video system, PC servers are used for video processing.

13.4.1.1 DVR-Based Installations

When using traditional CCTV systems, the surveillance video from analog cameras is fed into an IV-enabled DVR (Figure 13.3). DVRs have encoder cards that convert the video from analog to digital format and then perform the intelligent analysis (e.g., people counting or car license plate recognition). They also compress the video, record it, and distribute resulting alarms and video output to authorized operators.

In this architecture, each analog camera is connected by an individual coax cable to a DVR. DVRs are generally embedded devices. Some have proprietary video formats. Although this approach works adequately for small installations with a limited number of cameras, it is not scalable or flexible. Each DVR comes with a specific number of inputs, and adding even one additional camera entails the addition of another DVR, which is

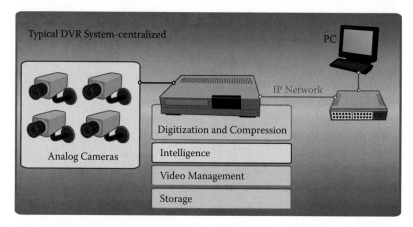

Figure 13.3 In a DVR-based installation, the IV functionality is located in the DVR along with all other functions such as digitization, compression, video management, and storage.

a costly proposition. In addition, because DVRs are proprietary embedded devices, they cannot be networked easily and do not support general network utilities such as firewalls and virus protection.

DVRs were traditionally designed to store and view a limited number of cameras and, as a result, they do not have much computational power. When DVRs run newer IV applications that require a lot of processing power, they can support only a fraction of the number of cameras they were designed to support.

13.4.1.2 PC Server–Based Installations

To overcome the limitations of DVRs, newer centralized architectures use commercial off-the-shelf (COTS) PC servers for video processing (Figure 13.4). The video from network cameras is brought directly to servers over a network. If the cameras are analog, the video is digitized first by video encoders and then transmitted over a network to a server.

This architecture is more flexible and scalable than proprietary DVR-based architectures because digitization and compression have been pushed out to the network cameras and video encoders. However, because the servers perform many of the processor-intensive tasks

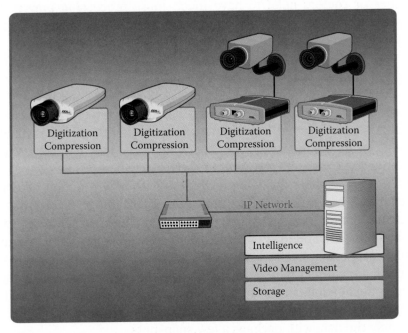

Figure 13.4 In a PC server–based installation, the IV functionality is located in the server along with all other functions such as video management and storage.

(transcoding the video, managing the storage, and processing the video for analysis), they need considerable processing power and each server is only able to process a relatively small number of cameras.

13.4.2 Distributed Systems

Distributed architectures are designed to overcome the limitations of a centralized system that overloads a central point such as a PC server or DVR. By distributing the processing to different elements in a network, bandwidth consumption can be reduced.

13.4.2.1 Network-Centric Installations

In typical network video systems, switches and routers are used to send video to appropriate components in a system. As a video stream passes through such gateways, the video data can be analyzed. The extracted metadata (see Chapter 13.5.1) can then be streamed instead of the video. This eliminates the dependency on a central unit and the potential bandwidth concerns in centralized infrastructures. However, because the switches or routers need to have much more processing power, their costs are higher. In addition, the design of the network is much more complicated (see Figure 13.5).

13.4.2.2 Intelligence at the Edge Installations

The most scalable, cost-effective, and flexible architecture is based on "intelligence at the edge," which means processing the video as much as possible inside the network cameras or video encoders. (Analog cameras do not have the capability to analyze video.)

Network cameras or video encoders with video motion detection, for example, can make use of this feature by sending video only when it detects motion in defined areas of a scene (Figure 13.6). Otherwise, no video is sent. The load on the infrastructure, including the required number of operators, falls dramatically. For specialized applications such as automatic number plate recognition or people counting, the impact of running applications in the camera is dramatic: the cameras can extract the required data (number plate information or number of people) and send only the data with perhaps a few snapshots.

This architecture uses the least amount of bandwidth because the cameras can send out metadata and intelligently figure out the required video to send. This significantly reduces the cost and complexity

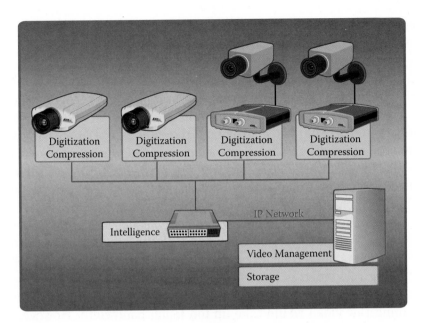

Figure 13.5 In a network-centric installation, the IV functionality is located in the network switches, not only making the system more scalable, but also rendering the design more complex.

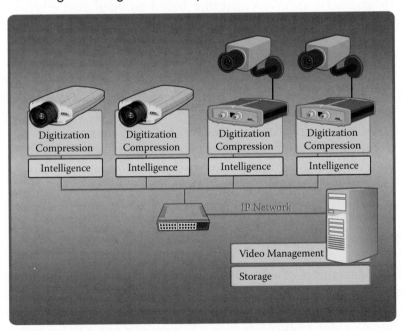

Figure 13.6 A truly distributed system with "intelligence at the edge" (i.e., in the network cameras and video encoders) provides the most scalable and cost-efficient system.

associated with a network-centric-based IV system and completely eliminates the drawbacks of a centralized architecture.

Another advantage of having video processing at the edge — or
partly at the edge — is that it significantly reduces the cost of the servers
needed in running IV applications. Servers that would typically process
four to twelve video streams in a centralized IV solution would be able to
handle more than 100 video streams if the intelligence processing were
done in the network cameras. In some applications, where simply the data
is needed and not the video — for example, people counting or automatic
number plate recognition — the resulting data can be sent directly into a
database, which further reduces the load on servers.

Processing video in intelligent edge devices also greatly enhances
the quality of the analysis because the cameras can process raw video
data before it is tainted by lossy compression formats such as MPEG-4.
Processing raw video is ideal for video-processing algorithms. In centralized architectures, compressed video is sent to the servers because
streaming raw video would take up too much bandwidth. However, the
servers would then have to decompress or transcode the compressed
video packets to process them. This increases costs by increasing the
number of servers required for a given number of cameras.

In summary, an IV architecture based on intelligent edge devices
can result in significant cost savings and enhanced performance by:

- *Reducing the costs of PC servers required to process the video.* Because
  of the need for fewer servers, there also are lower power consumption and maintenance costs. In addition, in certain environments
  such as retail stores where there are generally no "server rooms,"
  installing a large number of servers is simply impractical.
- *Reducing network bandwidth utilization and costs.* Reducing the
  data rates by streaming only essential information means that
  lower-priced network components can be used.

13.4.3 Integrating Intelligent Video Applications

Many manufacturers of video surveillance equipment, as well as providers
of video management software, supply IV functionalities with their products. These IV functionalities are often applicable for most installations —
for example, video motion detection. Occasionally, other IV functionalities
for specific applications are provided — for example, people counting.

Building robust and commercially viable IV applications with
a high level of functionality requires a great deal of expertise in areas

such as image analysis, which uses advanced mathematical algorithms. Sometimes, this requires specialized knowledge in a specific application area, such as retail, public transportation, or customs control. For this reason, a number of software companies have chosen to focus their skills on developing and supplying IV applications that solve specific needs in specific markets, allowing for tailored video surveillance solutions.

Although this provides greater freedom of choice for the end user, it also makes compatibility and easy integration necessary between network cameras, video management software, and the IV applications. To be commercially attractive and to optimize software compatibility and usability, cameras, software, and IV applications all must be built on open and published interfaces (application programming interfaces, or APIs) and platforms. This enables easy integration of the IV software directly into the cameras and also enables communication with video management systems.

From the point of view of IV suppliers, having platforms and standards provides the most flexibility and allows vendors to rapidly install their IV applications in cameras and systems from different manufacturers. By significantly lowering the cost and complexity of implementing intelligent video, it becomes possible for IV to rapidly penetrate the market, thereby enabling users to get the maximum benefit from their video systems at the lowest possible cost.

13.5 APIs and Standards

IV applications work on digitized streams of video, coming from either a network camera or a video encoder that has digitized analog video from an analog camera. For intelligent video applications to work as part of a video surveillance system, there must be standardized formats for digitized video streams. A number of video compression standards exist, some of which are more relevant for IV than others. These include Motion JPEG, MPEG-4, and MPEG-4 AVC/H.264. For more information about compression, see Chapter 5.

In addition to the requirement of having video compression standards, there is a need for open standards to describe and tag the content of video images. Information about the content of an image (or any content at all) is often referred to as "metadata," which is explained in more detail below.

13.5.1 Metadata

Metadata, which literally means "data about data," provides a solution to the challenge of sifting through volumes of recorded video to find, filter,

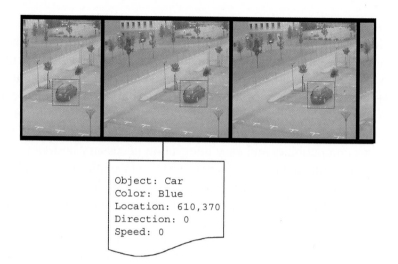

```
Object: Car
Color: Blue
Location: 610,370
Direction: 0
Speed: 0
```

Figure 13.7 A video stream tagged with metadata — for example, the object, color, and location of a car parked in a parking lot.

and retrieve the right information in a time-saving manner. Metadata allows for tagging of various contents of an image along with the image, and enables automatic analysis of video streams. This allows users to easily find exactly what they are looking for in a recording.

For example, an IV application that counts people passing through an area can tag a number to an image as it streams to a central server for further analysis. The IV application can even be designed to send only the required information — the number of people — and not stream images at all.

For video surveillance of a bridge or a highway, an IV application could potentially tag vehicles appearing in an image according to certain criteria and store the tags together with the video. This would make it possible for an operator to search for, for example, all red cars heading north in the past 24 hours and then instantly have access to the right video sequences, instead of having to manually watch 24 hours of recorded video to find all relevant occurrences (Figure 13.7).

Metadata based on open and standardized ways of describing information can be more easily integrated into various systems, allowing users to build open and scalable systems. A number of standards exist for metadata; many have been developed for specific areas such as libraries, geospatial use, and database management. For video surveillance and intelligent video, XML (eXtensible Markup Language) and potentially MPEG-7 are of particular interest.

13.5.2 XML and MPEG-7

MPEG-7 uses XML to store metadata, and it can be attached to a time code to tag particular events in a video stream. Although MPEG-7 is independent of the actual encoding technique of the multimedia, the representation that is defined within MPEG-4 (i.e., the representation of audio-visual data in terms of objects) is very well suited to the MPEG-7 standard.

One of the drawbacks of MPEG-7 is that it is a very extensive standard. If used for IV applications, only a very small portion of the standard is applied. At this time, MPEG-7 will most likely not be used extensively in IV due to its complexity. XML, on the other hand, is a good way to exchange data. There are also other data exchange formats and scene-descriptive languages that might emerge for IV applications in the future.

For products using any type of API, it is essential that the API is well documented, open, and backwards compatible as improvements are made.

13.6 Best Practices

Intelligent video is an emerging market with a wide variety of offerings from many companies. Athough the architecture of the solution is often overlooked, it is important to fully understand it to ensure that the solution is the right one. When choosing to implement IV, the solution should be based on open standards and able to scale as an installation grows.

Things to consider include:

- *Reliability of the system.* Is the architecture minimizing the risk of system failure and associated downtime?
- *Scalability and flexibility.* Can the system effortlessly scale from a few to many cameras? Can it intelligently distribute processing on different components of the network?
- *Interoperability.* Are the IV application and video management application tied to certain hardware, or can system components from different vendors be used?
- *Data format used by the IV application.* Is the IV metadata based on an openly published standard or API so that it can be easily incorporated into other systems?
- *Accuracy.* Remember that very few IV systems, if any, are 100 percent accurate. An operator can manage only a few false-positives a day.

Chapter 14

Intelligent Video Applications

A wide array of intelligent video (IV) applications is available today and more will become available in the coming years as the market emerges. Some applications such as video motion detection and camera tampering alarm are applicable for most video surveillance installations, whereas others address the needs of specific industries, mostly in the retail, transportation, and critical infrastructure markets.

This chapter provides an overview of the most common applications. They are categorized according to the technology used to provide the intelligence — that is, pixel based, object based, or specialized. Because IV is a fairly new technology, it is important to have realistic expectations. This is discussed at the end of the chapter, along with some best practices.

14.1 Categorizing IV Applications

Although IV applications can be categorized based on the industry sector in which they are used, a more straightforward approach is to categorize them according to the basic technology used — that is, whether it is pixel based, object based, or specialized.

14.1.1 Pixels, Blobs, and Objects

At a basic level, IV analysis software works by analyzing every pixel in every frame of video, characterizing those pixels, and then making decisions based on those characteristics. Basic IV applications make decisions

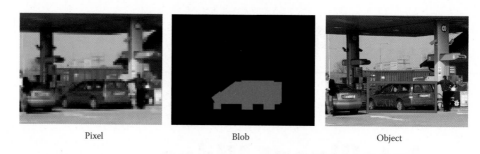

Pixel   Blob   Object

Figure 14.1 IV applications are based on detecting and tracking pixels, blobs, or objects.

based on changes in the characteristics of pixels (e.g., pixel-based motion detection), which can raise an alert when a certain number of pixels change based on criteria such as size, color, and brightness.

Blob recognition involves a level of intelligence beyond detection of pixel changes. A blob is essentially a collection of contiguous pixels that share particular characteristics; blobs have boundaries that delineate them from other parts of a video frame. Blobs can be analyzed and characterized as being particular objects. For example, a blob can be identified as a person or a car by analyzing its shape, size, speed, or other parameters. Applications based on object classification and tracking require the most sophisticated software algorithms. See Figure 14.1.

14.1.2 Categorization of Intelligent Video Applications

There are a large number of IV applications based on extracting different kinds of information from a video, processing it in different ways, and applying different rules for decision making. IV applications can be divided into three broad categories:

1. *Pixel-based IV applications.* These IV techniques are used to send an alert when a loss of video quality or detection of motion in images occurs. Examples of applications are:
   - Video motion detection
   - Camera tampering detection
2. *Object-based IV applications.* This is the category with the largest number of IV applications. It is based on the recognition and categorization of objects in an image. These applications fall into two broad categories:
   - Object recognition and classification
   - Object tracking

3. *Specialized IV applications.* These applications use both pixel- and object-based intelligence to process video for a specific application. Such applications include:
   - Number or license plate recognition
   - Facial recognition
   - Fire and smoke detection

The following sections discuss the different applications in detail.

14.2 Pixel-Based Intelligence

IV algorithms based on pixel analysis are very common. Many of the applications are applicable for most installations, making them widely useable. They include video motion detection, camera tampering detection, and image enhancement functionality, all of which are discussed in the following subsections.

14.2.1 Video Motion Detection

Video motion detection (VMD) is the original, most basic and prevalent IV application in video surveillance. It primarily is used to reduce the amount of video stored by flagging video that has some changes and ignoring video with no changes. By only storing video in which changes occur, security personnel can store video for a longer period on given storage capacity. It also is used to flag events (e.g., persons entering restricted areas) to operators for immediate action.

VMD is the foundation for a large number of more advanced IV applications, such as people counting, digital fences, and object tracking.

Software algorithms continually compare images from a video stream to detect changes in an image. Early IV systems that recognized motion in the camera view did so based on simply detecting pixel changes from one frame of video to the next. Although this scheme certainly helped reduce the amount of storage required (by not storing video in which nothing changes), it was not very useful for real-time applications because it gave too many false-positives. It raised too many alarms based on pixel changes caused by uninteresting events (e.g., minor light changes, slight camera motion, or motion of trees).

More advanced VMD systems have the intelligence to exclude pixel-based changes from known sources, such as natural changes in lighting conditions based on the time of day, or other known and repetitive changes in the camera's field of view. By excluding these, the number of false alarms drastically decreases.

Sophisticated IV systems deploy more advanced algorithms to not only detect individual changes in pixels but also group pixels together. In this way, the system can "understand" that many pixels together actually constitute larger objects, such as people or cars. This further reduces the amount of error in the motion detection.

A number of parameters are normally available in VMD applications for fine-tuning the system. These parameters include the threshold for how large an object should be for the system to trigger, a setting for how long the object should be moving in an image before it stops triggering the system, and a sensitivity for how much an image can change before the system reacts. Finding the right balance between these and other settings will directly impact the number of false alarms the system will give and whether all relevant motion in the scene is detected. For best results, the VMD should be fine-tuned when the cameras are installed, and it should be observed over a certain period of time to ensure a robust implementation. Advanced network cameras (Figure 14.2) often enable the placement of many different windows where motion will be detected. Each VMD window can then be defined separately according to different parameters.

Network video systems will often include a "pre-alarm buffer," which means that the system continuously records, for example, the last 30 or

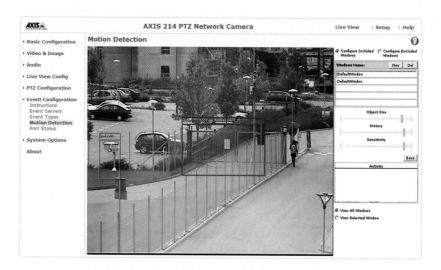

Figure 14.2 Advanced VMD functionality in network cameras includes the option of setting up several VMD windows, which then can be defined individually based on adjustable parameters such as object size, history, and sensitivity.

60 seconds. This ensures that when the VMD is triggered, the system will always have some pre-event video that can be sent along with video of an event. It usually is helpful for operators to see what preceded an event. A post-alarm buffer also is usually available for use and can be configured so that some seconds of video following an event are captured as well.

In analog CCTV (closed-circuit television) systems with DVRs (digital video recorders), cameras are connected to the DVR, which performs the VMD on each video stream. This allows the DVR to decrease the amount of recorded video in an effort to prioritize recordings. The downside of this setup is that performing VMD is a CPU (central processing unit)-intensive process, and performing VMD on many channels puts a heavy strain on the DVR system. The same downside exists in a network video system where the VMD functionality is performed by a video management software application on a server.

When "edge devices" such as network cameras and video encoders perform VMD, it saves bandwidth and storage, creates a much more scalable solution, and simplifies integration with other systems such as alarms. Chapter 13 provides further information on different IV architectures.

14.2.2 Camera Tampering Detection

Sometimes the views of cameras in a surveillance system can become obstructed or altered due to either intentional tampering or unintentional mistakes, making them unusable. For example, the lens of a camera might be deliberately or unintentionally covered (e.g., by paint, powder, moisture, or a sticker), or the cameras might be deliberately or unintentionally redirected to a view of no interest. More serious cases of tampering can include the actual removal of the cameras or severely defocusing them. Due to any of these reasons, surveillance cameras can become of limited use. It therefore is important to have the ability to automatically detect the tampering and raise an alarm. Otherwise, significant threats would go undetected, and completely unusable video would be stored.

Camera tampering detection is applicable in any installation but is used predominantly in environments that are potentially exposed to vandalism — such as schools, prisons, or public transportation — where it is likely that someone will intentionally redirect or block the cameras.

Imagine a video surveillance installation in a subway with thousands of cameras. Without camera tampering detection, a mistakenly redirected camera might be detected only after several months of pointing in the wrong direction, and most likely it will be detected when video from

Figure 14.3 Camera tampering detection will (within seconds) detect if a camera has been repositioned, is out of focus, or has been covered.

that specific camera needs to be reviewed, making that part of the system useless to the operator (Figures 14.3 and 14.4).

A tampering detection functionality must be capable of telling the difference between expected changes in a camera view versus unexpected changes due to tampering. Otherwise, false alarms would limit the program's usability. As with many IV applications, there are different ways of implementing tampering detection. It can be implemented in a camera or

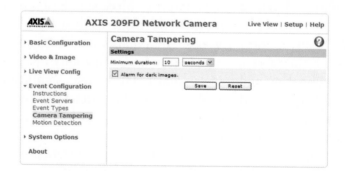

Figure 14.4 Some camera tampering detection will learn the scene automatically, making setup very easy. Simply select if camera tampering detection should be activated, along with the minimum duration, and the camera takes care of the rest.

in a centralized location such as with software on a server. By installing tampering detection algorithms in each camera, the system is more easily scalable compared with running the IV application in a central server.

14.2.3 Image Enhancement

Weather conditions naturally will affect the quality of video images in outdoor surveillance. Fog, smoke, rain, and snow will impact severely the possibility of monitoring scenes and identifying people, objects, or activities. In these situations, adjusting brightness and contrast of the images will only improve the situation slightly.

IV algorithms can be used to analyze video streams and detect typical distortions caused by bad weather (Figure 14.5). The same software then can, in real-time, enhance an image and restore it as much as

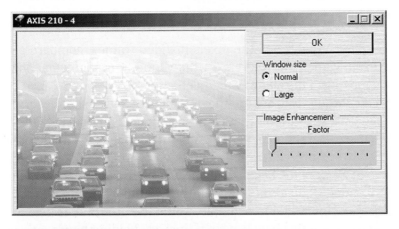

Figure 14.5 An image distorted by bad weather (top) and the image after image enhancement algorithms are applied (bottom).

possible to what it would have looked like without the distortion, resulting in a video stream with substantially improved image quality.

14.3 Object-Based Applications

Instead of detecting individual pixels, object-based applications are based on defining groups of pixels that define a certain type of object. When objects have been detected, metadata then can be extracted from the video based on detecting moving objects and characterizing them.

14.3.1 Object Classification

In general, most IV systems go through the following steps (Figure 14.6):

1. *Detection.* Analyze all the pixels in video frames, compare pixels in each frame to a "reference" frame, and figure out what objects are moving. This step is called detection.
2. *Segmentation.* Extract the moving objects and assign descriptive "signatures" to them, that is, descriptions using such criteria as color, size, direction of motion, and time. Together with detection, these steps are called segmentation. Segmentation is a critical step in IV systems because the accuracy and usefulness of the IV application depend on how well the system is able to extract the objects of interest. For example, would the program be able to tell that there are two people even if they are holding hands, or would they be interpreted as being one object?

 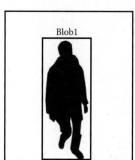 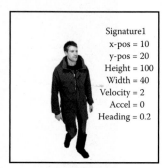

Figure 14.6 An object-based IV system will detect and segment a moving object first, and then classify it and provide a set of metadata that describes the object.

3. *Classification.* Once different objects have been segmented, they are classified into different object types (e.g., person, car) and assigned a set of descriptors that characterize the object based on, for example, color, size, or direction of motion. The descriptive information about the objects in an image is called metadata. Section 13.5.1 provides further information on metadata.

Once the metadata is available, a set of criteria can be applied — for example, a person walking the wrong way, a bag left behind by a person, or a car entering a restricted area. If the criteria are fulfilled, the system then can raise an alert in real-time or retrieve the appropriate video from storage.

For certain applications, the ability of IV systems to recognize the "type" of object greatly enhances their accuracy and usability. For example, rather than counting all objects in a scene, the IV system can recognize and count only particular objects, such as people or vehicles (Figure 14.7).

A specific challenge for IV systems involves the fact that objects can appear in different configurations than what is expected. For example, the system may be able to distinguish a human being from a dog, based partially on the knowledge that humans have a different aspect ratio, that is, they are substantially taller than they are wide. But if an individual is crawling, the perceived proportions of that person will substantially differ from the norm.

Figure 14.7 IV systems need to distinguish between interesting object types and non-interesting object types, as defined by the user.

A robust IV system needs to compensate for this — often called "aspect ratio compensation" — and be able to recognize a human being regardless of whether the person is crawling, crouching, standing, or running.

14.3.2 Object Counting

After detecting and classifying an object, the IV system can use the information for different purposes. One purpose is to count the number of objects that behave in a certain way, such as a person walking in the wrong direction. Another application can be to determine the total number of objects of a particular type in a scene.

14.3.2.1 People Counting

People counting is useful in environments such as retail stores where it is important to know the number of people entering or exiting an area. Although there are other technologies such as infrared that can count people, using video can in many instances provide better accuracy. When using an IV application for people counting, the placement of the camera becomes important (Figure 14.8). Ideally, camera placement should be immediately above the entrance. The height depends on the optic lens used and the width of the entrance. The size of the person passing under the camera must be larger than 6 percent of the camera's total horizontal field of view. The quality of the image from the camera must be good enough to be able to clearly distinguish the people passing under the camera.

The ability of IV systems to accurately count the number of people opens up a range of applications, including:

- *Customer traffic.* A retail store can use people counting to count the number of people entering and exiting the store, going through certain aisles, or stopping by a particular merchandising display (Figure 14.9).
- *Queue management.* To enhance customer service, people counting can be used to count the number of people standing in a queue for service. Examples where this application may be useful include queues at retail checkout counters and at airports for ticketing or passport and security control (Figure 14.10).
- *Tailgating.* This is an important access control application where the system can send an alert when multiple people enter a facility although only one person has swiped an access control badge to open the doors (Figure 14.11).

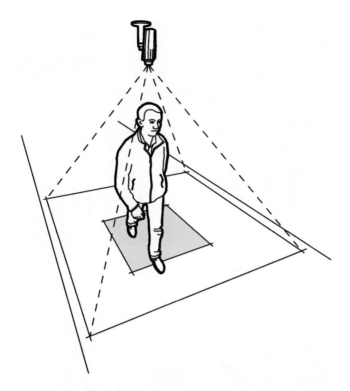

Figure 14.8 The placement of the camera becomes important in providing the appropriate accuracy.

Figure 14.9 An example of customer traffic application. (Picture courtesy of ClickIt, Inc.)

Figure 14.10 An example of queue management application. (Picture courtesy of iOmniscient Intelligent Surveillance.)

Figure 14.11 Example of a tailgating detection application. (Picture courtesy of Vidient Systems.)

14.3.2.2 Crowd Detection

Crowd detection is another application of people counting, whereby a program can count the number of people gathered in a particular area in a scene. Security personnel, for example, may want to be alerted when a crowd gathers in a particular area, or a security officer may want to know if there is a risk of overcrowding in a particular area in case of an evacuation. In these examples, the IV system counts the number of people in the given area and sends alerts if the number exceeds a predefined threshold (Figure 14.12).

Figure 14.12 An example of crowd detection. (Picture courtesy of iOmniscient Intelligent Surveillance.)

14.3.3 Object Tracking

In most security applications, it is necessary to keep track of where objects and people are in a facility. IV applications based on object tracking work by first segmenting a particular object in a camera view and then tracking the object as it moves around in that view or from one camera to another.

14.3.3.1 Digital Fence

This application is also called "tripwire" or "virtual fence." It is used to alert security personnel to possible access control breaches. Setting up such alerts usually involves designating a line or area and then telling the system to generate an alert if objects go past that line in a particular direction or if objects enter or leave a certain area (Figure 14.13).

An example might be someone trying to go through an exit door at an airport terminal. Another example might be a system set up to allow staff to leave a building after dark but to send an alarm if someone tries to enter it.

14.3.3.2 Object Left Behind

This is a critical application for the security of common areas. The application targets threats from explosives left behind in bags or packages.

Figure 14.13 Virtual fence can be used to detect and send an alert if someone enters the rail or subway track.

The IV application watches an area and keeps track of all objects in it. When an object that was previously moving becomes stationary and stays that way for a certain period of time, the system raises an alert and shows the operator the object of concern (Figure 14.14).

14.3.3.3 Loitering

A loitering functionality will track the amount of time and the number of people who linger in a certain area. People lingering or loitering — for example, in a parking lot or in front of a bank ATM (automatic teller machine) — may be an indication of malicious intent. The IV capability also is useful for retail merchandising applications — for example, in determining how long people stand in front of a merchandising shelf or how long people have been waiting in checkout lanes.

14.3.3.4 Single-Camera People Tracking

In this application, the IV system is capable of tracking a particular person moving within a camera's view. The person of interest can be selected by the operator or automatically selected by the system. The former is useful in crowded locations such as at airports and retail stores; the latter is used mainly for perimeter control applications where even the presence of a single person is of interest.

For the single-camera tracking application, it is advantageous to use cameras with a wide field of view; this will allow a security operator to

Figure 14.14 An example of a system for abandoned object detection. (Picture courtesy of Agent Video Intelligence, Inc.)

track the person of interest over a wider area. For example, some network cameras have a 140-degree field of view and will even allow an operator to zoom into a particular area without losing video quality. Although very wide view cameras, so-called 360-cameras, seem well suited to this type of tracking application, they are not very practical in security applications because they normally do not have high-enough resolution to provide the needed detail (Figure 14.15).

Figure 14.15 A people-tracking system will show the current location of a person and track where he or she came from. (Picture courtesy of Agent Video Intelligence, Inc.)

14.3.3.5 People Tracking Using PTZ Cameras

This is a special case of a single-camera tracking application where a PTZ (pan, tilt, zoom) camera is controlled automatically to keep the person in sight. The advantage here is that the camera can zoom automatically in to give a better view of the person. An issue here is that if only one PTZ camera is used, then it could be pointing in another direction than an event of interest. To address this, a hybrid approach can be used, wherein fixed cameras find the events, and then a PTZ camera is used for the tracking. Another issue with PTZ auto-tracking is that only one object can be tracked at a time, and some PTZ tracking systems get confused if more than one object is in the camera view.

There are two different types of automatic tracking:

1. *Automatic selection — automatic tracking.* In this scenario, the camera locks onto the first moving object until it loses that object. Then the camera will find another moving object. This solution is useful in low-traffic environments such as parking lots and hallways. It provides a view of the object without the need for personnel.
2. *Manual selection — automatic tracking.* In this scenario, the surveillance officer selects the object to track and the camera follows it. This setup helps the officer focus on the object instead of being distracted by operating the camera.

14.3.3.6 Multi-Camera People Tracking

Multi-camera people tracking, also called camera handoffs, is one of the most difficult IV applications. For locations and facilities that need to be covered by a large number of cameras, this application enables security personnel to constantly keep a particular suspect in view. To do this, the IV system must be able to "hand off" a particular object from one camera to another. In most facilities, cameras do not cover 100 percent of an area, so the IV system must remember the object that was being tracked even when it goes off a camera view and then must be able to recognize it again when it appears in another camera.

It is very difficult to accomplish fully automated multi-camera tracking. Some IV systems help security personnel track a suspect by giving them a narrow set of cameras where the suspect may appear as he or she goes off one camera. This narrows down the possible number of cameras on which the suspect can appear (Figure 14.16).

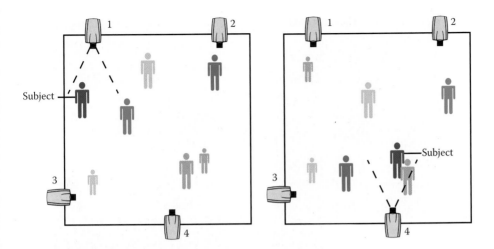

Figure 14.16 A multi-camera tracking system showing a person being tracked as he or she moves from one camera view to another.

14.4 Specialized Applications

Specialized applications are those designed to perform a particular task based on extracting specific features from a video and can be based on a combination of pixel- and object-based IV techniques.

14.4.1 Automatic Number Plate Recognition

Automatic number plate recognition (ANPR) often is referred to as automatic license plate recognition (ALPR, or just LPR). Automatically recognizing a car's number plate has a variety of uses, ranging from access control to looking for a particular vehicle. For example, in an access control application, only vehicles with particular number plates may have access to a facility. In the instance of a criminal investigation, ANPR can be used to automatically look for vehicles with a particular number plate that may have been involved in a crime.

In an intelligent parking application, ANPR can be used to automatically track vehicles coming into a parking lot, thus eliminating the need for a more expensive parking ticket infrastructure. ANPR also can be used to automatically monitor how long a particular vehicle stays in a parking lot. This can be applied, for example, by a store that wants to reserve its parking lot to shoppers only or receive alerts when a vehicle may have been abandoned.

In retail applications, ANPR can be used to identify cars whose occupants shop often in certain stores. The information then can be used, for

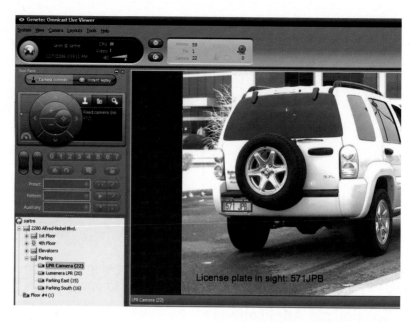

Figure 14.17 An example of an ANPR application. (Picture courtesy of Genetec Inc.)

example, to analyze where these shoppers come from or to design direct marketing programs that reach only consumers in the right geographic area. Figure 14.17 provides an example of an ANPR application.

An intelligent video application that performs ANPR applies a process that consists of several steps. In the car identification example, for instance,

1. If the car is not parked in a known, well-defined physical location (such as the gate to a parking house), the first step consists of finding the car in an image. For cars in motion, video motion detection is a key element in making this work. For parked cars, it is a matter of recognizing the outline of a car in an image.
2. Once the car is found, the next step is to isolate the actual number plate from the rest of the image, based on algorithms that define what a license plate looks like and where it might be mounted on a vehicle.
3. The next step consists of extracting the letters and the numbers from the license plate using image analysis.
4. Optical character recognition (OCR) is applied. The characters are transformed from a collection of pixels into a stream of letters and numbers.
5. The final step consists of processing the resulting string of letters and numbers by storing it in a database or comparing it with existing entries.

Figure 14.18 In an ANPR application, specialty cameras are often used to ensure that a clear capture of the number plate can be provided at all times of the day. A second camera can be used to provide an overview image of the car. (Picture courtesy of Extreme CCTV.)

It is especially difficult to deal with some challenges. Bad weather, light shining from headlights, and dirty or bent license plates can affect the result of the process. In addition, license plates look different in different parts of the world, which means that the IV application should be adapted to local conditions and quite often fine-tuned to the specific implementation. Specialty cameras are often used to provide the best images possible for the ANPR application (Figure 14.18).

From an implementation design point of view, ANPR is an IV application where the benefits of a distributed approach are very clear. Deploying an ANPR application in a network camera enables, if desired, the transmission of only the letters and numbers of a license plate to a central location — with perhaps just a snapshot of the vehicle — which drastically reduces the network load compared with a centralized implementation of an ANPR system.

14.4.2 Facial Recognition

Facial recognition is one of the more high-profile possibilities with intelligent video. The application areas are practically endless in number. Police are interested in getting alerts when certain individuals are seen in public spaces or in sensitive areas. Allowing only certain people to enter specific areas can enhance access control. For forensic purposes, the need can arise when searching for individuals in stored videos. In casinos, owners may want to catch blacklisted individuals. Customs can improve the precision of border control by augmenting manual checking of passports with automatic searches for certain individuals.

The process for a facial recognition IV system is similar to the automatic number plate recognition system described above. One difference is that for a number plate, it is possible to define beforehand what the system should look for — that is, a string of letters and digits grouped in a certain order. Most facial recognition systems are designed to do just that: recognize a face, which means that they depend on the initial step of actually building a database of "wanted" faces from, for example, a passport photo database or a police register. Sometimes getting that data becomes the biggest challenge in a facial recognition system.

Beyond that, the steps are as follows:

1. If the person in question is not standing still in a predefined location, the first step consists of finding the person in an image.
2. The next step is often called "face finding," which means isolating the face from the rest of the body.
3. The subsequent step involves identifying typical parts of the face, in terms of locating the position and recognizing the shape of features such as the eyes, nose, mouth, chin, skin color, and hair. From these traits, a unique pattern of the individual's face is constructed. In some applications, face finding is sufficient. For example, in an airport, it can be used to measure the queue time from entering and exiting a check-in point. In this case, the actual identity of the individual is not interesting, but the system should be able to separate different individuals from each other. Another application area can be in retail, where the system is calculating the "conversion ratio," that is, how many customers, out of the total number of people who entered the store, made a purchase. Here, the system needs to distinguish individual faces, but it is not relevant who they actually are. In a monitored situation, it may be enough for a guard on duty to be presented with individual faces and to match them with registered faces from access control badges.

Figure 14.19 An example of a facial recognition system. (Picture courtesy of Vigilant Video.)

4. The final step involves matching the extracted face with signatures from a database to actually identify individuals.

As with ANPR, the challenges for facial recognition systems are substantial. Even under perfect lighting conditions, people generally move around and may block each other, and appearances change over time or can be easily changed with, for example, glasses, hair color and length, or beards. A particular challenge involves the fact that people who are moving about freely will very rarely look straight into a camera. To deal with this, so-called three-dimensional (3D) facial recognition systems have been introduced. They work by extracting three-dimensional information from video streams and matching it with a database. Figure 14.19 provides an example of a facial recognition system.

14.4.3 Fire and Smoke Detection

Traditional smoke detectors trigger when there is a sufficient amount of airborne smoke in a room. Fire and smoke detection as part of IV applications will process video streams and look for visual cues to fire and smoke being present, which gives the advantage of being able to potentially react faster than a smoke detector. An IV-based fire and smoke detector may be able to react as soon as flames are visible in a room, without having

Figure 14.20 An illustration of privacy masking where only the provision of the right password can undo the masking. (Picture courtesy of EMITALL Surveillance SA.)

to "wait" until there is smoke or until the smoke has reached the ceiling. Such an application may be a good complement to traditional types of smoke detectors.

The IV system will process video images and react when combinations of color, light, and movement that are typical for fire or smoke appear. An alarm then can be sent together with relevant video images to an alarm central.

14.5 IV and Privacy Issues

IV applications such as facial recognition or people counting are sometimes seen as being invasive of people's privacy. However, users of such applications can easily overcome these concerns by not storing actual pictures or videos of the faces. For example, in a people-counting application, once the data is collected, there is no reason to keep the picture or video of individuals.

In some ways, IV applications can enhance privacy rather than the opposite. For example, an IV application may be able to find and mask out all the people in a video surveillance recording of a public area and then allow only the police to "unlock" these images if required in an investigation (Figure 14.20).

14.6 Realistic Expectations of Intelligent Video

This chapter presented an overview of a variety of IV applications, from the more generic (such as video motion detection) to very advanced systems (such as general facial recognition systems). It is important to keep in mind that intelligent video is still a very new field of knowledge and that many of the algorithms are still in their infancy.

Although all the applications described above indeed can perform well in a controlled environment, many will struggle to be robust and efficient enough to actually function optimally in real-life situations. Some of the IV systems will work only after tremendous efforts have been made to tailor the system to every single detail in a certain application.

Caught up in the enthusiasm of what IV could do, numerous vendors and software suppliers in the early twenty-first century created too high expectations in the video surveillance market for what is actually possible or cost effective. Although the market has calmed down somewhat recently, IV still struggles with unrealistic expectations in many areas.

Generally, the IV applications that enhance traditional camera functionality — such as video motion detection, image enhancement, and camera tampering — have matured well and now can be used in most applications without major adaptations or customizations. Other applications, such as people counting, ANPR, and digital fences, also are starting to be used more broadly, although they typically will require a fair amount of calibration in each installation. Some of the most advanced IV applications should still be considered demonstration software rather than real-world applications that will work for the majority of users.

14.7 Best Practices

When considering an IV application, four important technical factors must be present for intelligent video to work accurately:

1. The right input quality (i.e., video image quality)
2. Efficient IV algorithms
3. Ample computer processing power
4. Configuration and fine-tuning

14.7.1 Video Image Quality

One of the major considerations in making IV applications work well is getting the right video from the cameras, that is, video that is suitable for processing in a particular application. Some important considerations include:

- *Frame rates and resolution.* Contrary to popular belief, most IV applications do not need high frame rates and high resolutions. In fact, five to ten frames per second in CIF (equal to circa 0.1 megapixel) resolution is optimal for most general intelligence applications. Some specialized applications, such as ANPR and facial recognition, may require even lower frame rates, although they work better at higher resolutions.
- *Camera position.* Positioning cameras for IV applications is critical to getting accurate results. For example, most people-counting applications work best with cameras positioned overhead, which allows IV algorithms to separate individuals most effectively. For applications like these, it is best to dedicate particular cameras because an overhead camera is not optimal for recognizing people in security applications. Applications such as ANPR and facial recognition work best with frontal views, that is, with the camera looking straight at a number plate or face.
- *Type of video.* For most IV software, it is optimal to process uncompressed video because compression always results in loss of data. Some IV applications that require object classification need color video; others that simply count or track can use black-and-white video. A very interesting possibility is running IV based on data from thermal cameras. Although this is still premature given the high cost of thermal cameras, thermal cameras are extremely well suited to IV applications that require precision in telling people apart.

14.7.2 Efficient Intelligent Video Algorithms

IV applications are built on complex mathematical algorithms that process video and still images, each consisting of a myriad of details. The quality of an IV application will depend on how accurately the algorithm performs these calculations and how robust it is in dealing with variations in the input (i.e., the video stream). The only sure way of assessing the quality of an IV algorithm is to field-test the application under realistic conditions and see how fast it is and how many correct responses and false alarms it generates.

In general, 90 percent accuracy is achievable for a modern IV system. Reaching 95 percent, however, is very complex, and 99 percent or beyond is extremely difficult in a real-world situation. From a user's point of view, the demands for accuracy will differ. The demands can be influenced by, for example, how critical the IV application is for the safety and security of people and property, how many errors are acceptable, how many false alarms the system can be allowed to generate before it is unusable, and how many missed "true-positives" (i.e., situations that should have generated alarms but failed to do so) are acceptable. In addition, the cost of an IV system must be weighed against other alternatives such as having more security personnel.

14.7.3 Computer Processing Power

Because IV applications are mathematically complex and therefore computer power intensive, the performance will depend on the processors used and the amount of memory available. Some IV applications are optimized to run on small, embedded systems and will perform well in distributed architectures, whereas others demand a very powerful centralized server to perform reliably. In either case, the more processing power available to an IV application, the faster it will be able to perform.

14.7.4 Configuration and Fine-Tuning the System

IV applications and algorithms are designed to handle a large variety of situations. Every installation should be configured and fine-tuned for each particular scenario. No system is perfect, and configuring a system is a balance between not missing essential situations and reducing false triggers. Optimizing a system can take anywhere from a day up to weeks.

As a general rule of thumb:

- The system will never be 100 percent accurate.
- The more parameters that can be adjusted in an application, the longer it will take to optimize.
- Monitor and adjust the configuration over a 24-hour period, as changes in lighting will have an impact on the result.

Chapter 15

System Design Considerations

One of the main benefits of a network video system is the ability to mix and match and select the most appropriate components from different vendors. In addition, the system can be scaled to virtually any size. This flexibility and scalability mean that good knowledge about all the different components and how they interact is required. Selecting the right camera and properly installing and protecting it are, therefore, essential. Meanwhile, the appropriate network and storage solutions depend greatly on the selected cameras, the camera settings (e.g., resolution, compression, and frame rate) and the number of cameras.

This chapter discusses the most important aspects of designing and installing a network video system: how to select, install, and protect a network camera, and how to calculate the storage and network bandwidth. A design tool, such as the one on the DVD supplied with this book, is referenced. Some legal aspects of video surveillance are cited at the end of the chapter.

15.1 Selecting a Network Camera

In quickly growing markets such as the network video market, many new vendors are entering with new products. There are today more than 200 different network camera vendors in the marketplace. Because network cameras include much more functionality than analog cameras, choosing the right camera becomes not only more important but also more difficult. This section outlines the considerations to keep in mind when selecting a network camera. They include the type of camera, image quality, resolution, compression, and networking and other functionalities, as well as the vendor.

15.1.1 Type of Camera

To determine the type of network cameras required, as well as the number of cameras necessary to adequately cover an area, the scene or environment and the goal of the surveillance application must be determined first. Considerations include:

- *Indoor or outdoor camera.* If the camera will be placed outdoors, it should be installed in an appropriate protective housing, and a camera with an auto iris functionality is recommended.
- *PTZ or fixed camera.* PTZ (pan, tilt, zoom) or dome cameras with high optical zoom capabilities can provide highly detailed images and survey a large area. Keep in mind that to make full use of the capabilities of a PTZ or dome camera, an operator is required or an automatic tour must be set up. For surveillance recordings with no live monitoring, fixed network cameras are normally more cost-effective.
- *Light sensitivity and lighting requirements.* Consider additional lighting, specialized lighting such as IR (infrared) lamps, and light sensitivity levels of a camera. Day–night functionality can be a consideration. The light sensitivity of a camera should be evaluated. Do not go by the measurements on a datasheet, as vendors measure in different ways. For more information on light sensitivity measurements, see Section 4.1.2.
- *Tamper- or vandal-proof and other special housing requirements.* Proper protection against water, dust, temperature, and vandalism is essential. See Section 15.3 for more information.
- *Overt or covert surveillance.* This will help in selecting cameras that offer a non-discreet or discreet installation.
- *Area of coverage.* For a given location, determine the number of interest areas, how much of these areas should be covered, and whether the areas are located relatively close to each other or spread far apart. For example, if there are two relatively small areas of interest that are close to each other, a megapixel camera with a wide-angle lens can be used instead of two non-megapixel cameras.

- *Overview or high-detail images.* Determine also the field of view or the kind of image that will be captured: overview (e.g., to view a scene in general or look at the general movements of people) or high detail for identification of persons or objects (e.g., face or license plate recognition, point-of-sales monitoring).

15.1.2 Image Quality

Although image quality is one of the most important aspects of any camera, it is also something that cannot be quantified and measured, thus making it very difficult to choose the right camera. To illustrate the challenge, consider the two images in Figure 15.1, which were taken from two different cameras and under the same conditions, with lighting about 10 lux. Both cameras are similarly priced and from brand-name vendors. The conclusion is that the best way to determine image quality is to install different cameras and look at the video. In addition, keep in mind that although a camera may provide good still images, the images may not be as good when a lot of motion is introduced into the scene.

Progressive scan may be an important feature to have in a network camera, particularly if the scene to be captured involves high movement. Progressive scan consistently produces the best results in clarity and in providing the ability to recognize important details. Although progressive scan capability is found only in network cameras, not all network cameras have this functionality. For more details about progressive scan, see Section 4.4.3.

15.1.3 Resolution

In the video surveillance industry, some best practices have emerged regarding the number of pixels required for certain applications. For an overview image, it is generally considered that about 70 to 100 pixels are enough to represent 1 meter (20 to 30 pixels per foot) of a scene. For applications that require detailed images, such as face recognition, the demands can increase to as many as 500 pixels per meter (150 pixels per foot). This means that if there is a requirement to positively identify people passing through an area that is 2 meters wide by 2 meters high (7 by 7 feet), the camera needs to provide a resolution of at least 1 megapixel (1,000 by 1,000 pixels).

(a)

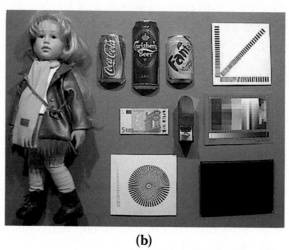

(b)

Figure 15.1 Two images taken under the same lighting condition using two similarly priced cameras can provide very different image quality. Notice the high amount of noise in image (a).

15.1.3.1 Determining the Resolution Needed

The required resolution for video surveillance images depends on the size of and the distance to the objects under surveillance. To illustrate this, consider a surveillance application at an airport entrance where there is a need to identify people in case of an incident.

The entrance area may be 20 meters (60 feet) wide and the aim is to be able to identify people. The estimated vertical field of view needed is 2 meters (6 feet). Figure 15.2 shows images of a face using four different resolutions, measured in number of pixels across the face. It is a

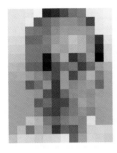

Figure 15.2 The effect of using increasing resolution. In the images above, 8, 16, 32, and 64 pixels across the face are used.

convenient measurement because the width of a face is about the same for all adults (about 0.15 meters, or 6 inches). The alternatives that are presented in the last two images, i.e., at 32 or preferably 64 pixels, would be required to positively identify a person.

By combining the needed resolution and the size of the surveillance area (20 by 2 meters using the airport entrance scenario mentioned previously), one can calculate the total number of pixels needed to cover the scene. Using the resolution alternatives shown in Figure 15.2, one can clearly see the impact of higher resolutions (see also Tables 15.1 and 15.2).

With the resolutions determined, the number of required cameras can be calculated. The possible camera combinations and the benefit of megapixel cameras can also be shown (Table 15.2).

Although somewhat hypothetical, the examples in Table 15.2 show that to achieve the highest resolution, the cost in terms of the number of cameras required is quite high. If the plan is to cover this scene with resolution Alternative 4, 28 VGA cameras are required. The number is

Table 15.1 Required Number of Pixels per Scene Given That the Distance across a Typical Face is 0.15 meters (6 inches)

Alternative	Resolution (pixels across face)	Scene (m)	Scene (pixels)
1	8	20 × 5	1070 × 110
2	16	20 × 5	2130 × 215
3	32	20 × 5	4270 × 430
4	64	20 × 5	8530 × 860

Note: For example, in the first line, one makes the following calculation to get the number of pixels horizontally: 20/0.15 × 8 = 1067 (or about 1070).

Table 15.2 Number of Cameras Needed to Cover the Scene

Alternative	Scene (pixels)	VGA (640x480)	1.3 Mpixel (1280x1024)	2 Mpixel (1600x1200)
1	1070 × 110	2 × 1 = 2	1 × 1 = 1	1 × 1 = 1
2	2130 × 215	4 × 1 = 4	2 × 1 = 2	2 × 1 = 2
3	4270 × 430	7 × 1 = 7	4 × 1 = 4	3 × 1 = 3
4	8530 × 860	14 × 2 = 28	7 × 1 = 7	6 × 1 = 6

reduced to six cameras if 2-megapixel cameras are used instead — a number that is much easier to handle.

15.1.4 Compression

Two of the most common types of video compressions are MPEG-4 and Motion JPEG. A relatively new compression technique being made available on network cameras is H.264. Each of the video compression techniques employs different ways of reducing the amount of data transferred and stored in a network video system while at the same time maintaining a level of image quality. Each format has its advantages and disadvantages. When network cameras offer support for more than one type of video compression standard, they give users greater flexibility in optimizing viewing and recording needs; for example, some systems may want to use MPEG-4 for live viewing and Motion JPEG for recording.

Designing the network as well as the storage system largely depends on the selected compression standard. The following are some basic recommendations:

- *Motion JPEG:* suitable for smaller systems with limited retention requirements, systems with frame rates less than 5 fps, and systems with remote cameras and uncertain bandwidth.
- *MPEG-4:* suitable for mid-sized and large systems. The higher the frame rates and the longer the retention times, the more beneficial MPEG-4 becomes. It also is ideal for systems with limited but stable bandwidth.
- *H.264:* Today there is a limited selection when it comes to products supporting H.264, but that is likely to change quickly. H.264 has all the benefits of today's MPEG-4 compression, in addition to the fact that it is twice as efficient, which enables greater savings in bandwidth and storage.

For more information on compression, see Chapter 5.

15.1.5 Networking Functionality

In the same way that high image quality is essential, a camera's networking functionality is just as important. Plugging into an Ethernet connection with an IP address is a basic functionality; all network cameras can boast the same. Several other technologies also should be supported in a professional network camera:

- *Power over Ethernet (PoE).* When a network camera supports this feature, it means that the camera can receive power through the same cable as for data. PoE reduces cabling requirements and installation costs and can provide savings of hundreds of dollars per camera. Make sure the camera's PoE support is in accordance with the IEEE 802.3af standard. This will give users the freedom to select from a wide variety of network switches. For more on PoE, see Section 8.6.
- *Extensive networking functionality and security.* Some examples of important network and network security functionalities include:
  - DHCP, used by many organizations to manage IP addresses
  - HTTPS encryption for secure communication
  - IP address filtering, which enables only defined IP addresses to have access rights to the camera
  - 802.1X for port-based authentication
  - IPv6, which supports the next IP addressing standard

  An important litmus test is the opinion of the IT department. It should be able to determine if the camera provides adequate networking functionality and security. For more on networking technologies, see Chapter 10.
- *Wireless technology.* Wireless is a good option when running a cable between a LAN and a network camera is impractical, difficult, or expensive. Wireless access also can be made to a standard network camera, but it must be connected to a wireless device point. Wireless technology can be useful, for example, in historic buildings where the installation of cables would damage the interior or within facilities where there is a need to move cameras to new locations on a regular basis, such as in a supermarket or in outdoor installations. Wireless technology also can be used to bridge sites without expensive ground cabling. For more on wireless networks, see Chapter 9.

15.1.6 Other Functionalities

- *Audio* enables users to remotely listen in on an area and communicate instructions, orders, or requests to visitors or intruders. Audio also can be used as an independent detection method. When detecting sound above a certain level, video recordings and alarms can be triggered. Consider whether one-way or two-way audio is required. A network camera with audio support comes either with a built-in microphone or an input for an external microphone. Speakers can be built in or external. See Chapter 6 for more on audio.

- *Built-in intelligence and analytics.* Certain network cameras may offer such built-in intelligence as video motion detection, tampering detection, and people counting. When such intelligence is located at the camera, it becomes scalable and can help reduce bandwidth and storage requirements because the camera is able to decide when to send and process video. Video intelligence requires a large amount of processing power; if that intelligence is not in the camera but rather on a PC server, which must process lots of data from several cameras, the PC can quickly become overloaded. See Chapters 13 and 14 for more on intelligent video.

- *Input/output (I/O) connectors* enable connection of external devices to a network camera. Inputs to a camera (e.g., a door contact, infrared motion detector, glass break sensor, or shock sensor) enable the camera to react to an external event by, for example, initiating the sending and recording of video. In the case of an application where the goal is to capture the identity of a person at an entrance, there is no need for the camera to continually send video. Using the input port, the camera can be triggered to capture and send the necessary image frames only when the door opens. Outputs enable the camera or a remote operator to control external devices by, for example, activating alarms, triggering door locks, generating smoke, or turning on lights.

- *Alarm management functions.* Advanced network cameras can process and link input, output, and other events. This is called alarm management. Pre- and post-alarm image buffers within a network camera can save and send images collected before and after an alarm occurs. After detecting an alarm or event, a network camera can send notifications via e-mail, TCP, and HTTP and upload images via FTP and HTTP. For example, if the level of audio in a room passes a preset threshold, the lights in the room can be turned on via the camera output, and video can be sent to the video management software.

15.1.7 Vendor

With more than 200 network cameras on the market, it is important to consider the following when selecting a network camera vendor:

- *Wide product portfolio.* When choosing network video vendors, go with those who maintain a full product line, including fixed cameras, fixed dome cameras, and PTZ dome cameras. This way, one or two companies can satisfy the needs now and well into the future when the system is expanded and upgraded with functionality such as megapixel, wireless, or audio. If an analog system must be integrated into a network video system, make sure that the chosen company's product portfolio also includes video encoders and decoders.
- *Extensive application support and ease of integration.* Make sure to select network cameras that have open interfaces (an application programming interface, or API) and are integrated with multiple video management software applications. Some network camera vendors have hundreds of such alliances. Open, multi-vendor video management systems give users the most flexibility.
- *Tools for managing large deployments.* Like all intelligent network devices, network cameras have an IP address and built-in firmware, enabling easy upgrades. Many vendors provide upgrades free of charge. When making a purchase decision, consider the cost to set IP addresses and eventually update all the cameras in the facility. The network camera vendor should have tools to manage these processes, and their estimates for cost and downtime should be clear and measurable up front. The vendor also should have programs that have the capability to automatically locate all network video devices and monitor the status of those devices.
- *Appropriate warranty.* Video surveillance systems are a substantial investment for most organizations and have a life expectancy of several years. Make sure the vendor has the appropriate warranty coverage on the selected cameras.
- *Select a vendor that will be a long-term partner.* Does the company have a large installed base of cameras, focus on network camera technology, and offer local representation and support? Is the company a global player? Because needs change and grow, it is

important to choose cameras from a vendor where the innovation, support, upgrades, and product path will be there for the long term. It is important to make network camera purchases based on the assumption of future growth and the need for added features and functionality. This means the network camera manufacturer should be a long-term partner. Do not sacrifice future security just to save a little money up front.

After making a decision as to the desired camera, it is a good idea to purchase one and test its quality before setting out to order quantities of it.

15.2 Installing a Network Camera

After purchasing a network camera, the way it is installed is just as important. Below are some recommendations on how to best achieve high-quality video surveillance based on camera positioning and environmental considerations.

15.2.1 Surveillance Objective

When positioning a camera, it is important to keep in mind the kind of image that one wants to capture. If the aim is to get an overview of an area in an effort to track the movement of people or objects, make sure a suitable camera is used and that it is placed in a position that achieves the objective.

If the intention is to identify a person or object, the camera must be positioned or focused in a way that will capture the level of detail needed for identification purposes. It may be favorable to place a camera in a high position to limit tampering. However, a lower placement may improve identification of faces or detailed objects, avoiding a "bird's-eye" perspective. Local police authorities also may be able to provide guidelines on how best to position a camera. A spinning Rotakin (see Figure 15.3) also can be used to test how well a camera displays moving objects.

15.2.2 Use Lots of Light or Add Light If Needed

The most common reason for poor quality images is lack of light. Generally, the more light, the better the image. It is normally easy and cost effective

Figure 15.3 Spinning Rotakin.

to add strong lamps in both indoor and outdoor situations to provide the necessary light conditions for capturing good images.

Table 15.3 shows the available light in different kinds of conditions. At least 200 lux is needed to capture good-quality images. A high-quality camera may be specified to work down to 1 lux, which means that it can capture an image at 1 lux, but it may not be of high quality. Different manufacturers use different references when specifying the light sensitivity of a camera, and this makes it difficult to compare cameras without first testing them and comparing the images captured.

When using external artificial lighting in outdoor environments, reflections and shadows should be avoided.

Table 15.3 Typical Illuminance on Surfaces in Different Lighting Situations

- Direct sunlight: 100,000 lux
- Open shade: 10,000 lux
- Overcast daylight: 1,000 lux
- Full moon light: 0.1 lux
- Sunrise or sunset: 400 lux
- Poorly lit room: 100 lux

For covert security or in areas where the presence of artificial light is unwanted, choose an IR-sensitive, black-and-white camera or an automatic, day/night camera. In a day/night camera, color video is delivered during daylight conditions, whereas at night the camera makes use of reflected near-infrared light to deliver black-and-white images. An IR illuminator, which provides near-infrared light, also can be used in conjunction with an IR-sensitive camera or a day/night camera to further enhance the camera's ability to produce high-quality video in low-light or nighttime conditions. For more information on IR illuminators, see Chapter 3, Section 3.5.1.

15.2.3 Avoid Direct Sunlight

Always avoid direct sunlight into the camera. Direct sunlight will "blind" the camera and can reduce the performance of the image sensor chip. If possible, position the camera with the sun shining from behind the camera.

15.2.4 Avoid Backlight

Avoid bright areas in an image. These areas may become overexposed (bright white) while objects in front can appear too dark. This problem typically occurs when attempting to capture an object in front of a window. To solve this problem, simply reposition the camera or use curtains and close blinds if possible. If it is not possible to reposition the camera, add frontal lighting. Cameras with support for a high (or wide) dynamic range are better at handling a backlight scenario (Figure 15.4).

- *Reduce the dynamic range of the scene.* In outdoor environments, viewing too much of the sky results in too high a dynamic range. The camera will adjust in order to achieve a proper light level for the sky. Consequently, the object or landscape of interest will appear too dark. One way to solve this problem is to mount the camera high above the ground, using a pole if needed.
- *Adjust camera settings.* It may be necessary at times to adjust settings for brightness, sharpness, and white balance for different environments (indoor, outdoor, fluorescent) to obtain an optimal image (Figure 15.5).

When deciding upon the exposure, a shorter exposure time is recommended when rapid movement occurs or when a high frame rate is

Figure 15.4 (Left) Avoid very bright areas in an image by changing the camera position. (Right) Advanced cameras include a feature that compensates for backlight.

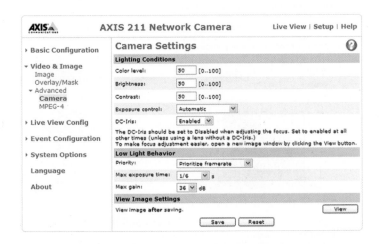

Figure 15.5 Example of a camera user interface showing options for advanced camera settings.

required. A longer exposure time will improve image quality in poor lighting conditions, but it may lower the total frame rate and result in increased motion blur. In some network cameras, an automatic exposure setting means the frame rate will increase or decrease with the amount of available light. It is only as the light level decreases that artificial light or prioritized frame rate or image quality becomes an important consideration.

15.2.5 Lens Selection

An auto iris lens should be used for outdoor applications. This lens automatically adjusts the amount of light that reaches the image sensor. This optimizes the image quality and protects the image sensor from being damaged by strong sunlight. The fastest way to calculate the lens required and the field of view is to use a rotating lens calculator (Figure 4.14), which is available from most camera and lens manufacturers. Several manufacturers also make their lens calculators available online (Figure 15.6), making the calculation quick and convenient. For more information on lenses, see Section 4.2.

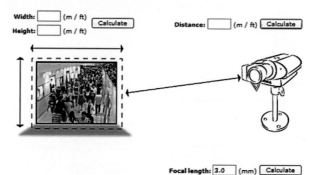

Lens calculator: AXIS 211, AXIS 211A

The focal length of the delivered varifocal lens is 3.0-8.0 mm. Keep this value or enter the focal length of the new lens to find out at what distance you should place the camera in order to capture a specific scene. You can also calculate the focal length of the lens you need by specifying the actual distance and scene dimensions.

Please enter the length values (distance, width or height) specified in the same measurement unit, e.g. meters or feet. The focal length of the lens is always specified in millimeters. Click the button to the right of the parameter you want to calculate.

Width: [ ] (m / ft)   [Calculate]   **Distance:** [ ] (m / ft) [Calculate]
Height: [ ] (m / ft)

Focal length: [3.0] (mm) [Calculate]
Example: If the focal length is 3.0 mm and the distance to the scene is 8 meters, the calculated width and height of the scene will be 9.60 and 7.20 meters, respectively.

Figure 15.6 Example of an online lens calculator.

15.3 Protecting a Network Camera

Surveillance cameras often are placed in very demanding environments. For outdoor installations, protection against varying weather conditions is necessary. In industrial settings, cameras may require protection from hazards such as dust, acids, or corrosive substances. In vehicles such as buses and trains, cameras must withstand high humidity, dust, and vibrations. Cameras also may require protection from vandalism and tampering.

Manufacturers of cameras and camera accessories employ various methods to meet such environmental challenges. Solutions include placing cameras in separate, protective housings; designing built-in special-purpose camera enclosures; and using intelligent algorithms that can detect and alert users of a change in a camera's operating conditions.

The level of protection provided by enclosures — whether built in or separate from a camera — is often indicated by classifications set by such standards as the IP, NEMA, and IK ratings.

The following subsections discuss such topics as coverings, positioning of fixed cameras in enclosures, environmental protection, vandal and tampering protection, types of mountings, and protection ratings.

15.3.1 Camera Enclosures in General

A camera's operating conditions are defined based on the materials and components used. When the demands of the environment are beyond a camera's operating conditions, protective housings are required. Although there are cameras such as fixed domes with built-in, specially designed casings, most enclosures are separate from the camera itself.

Camera housings come in different sizes and qualities, and with different features. Housings are made of either metal or plastic and can be generally classified into two types: (1) fixed camera housings and (2) dome camera housings. When selecting the appropriate enclosure for a specific camera, several things must be considered, including:

- Side or slide opening (for fixed camera housings)
- Mounting brackets
- Clear or smoked bubble (for dome camera housings)
- Cable management
- Temperature and other ratings (consider the need for heater, sunshield, fan, and wipers)
- Power supply (12, 24, 110 volts, etc.)
- Level of vandal resistance

Figure 15.7 (Left) Example of a housing with a built-in antenna; and (right) a bundle with an external antenna installed some distance away from the camera housing.

Some housings also have peripherals such as antennas for wireless applications (Figure 15.7). An external antenna is only required if the housing is made of metal. A wireless camera inside a plastic housing will work without the use of an external antenna.

15.3.2 Transparent Covering

The "window" or transparent covering of an enclosure is usually made of high-quality glass or durable polycarbonate plastic. Because windows act like optical lenses, they should be of high quality to minimize their effect on image quality. When there are built-in imperfections in the clear material, clarity — and therefore image quality — is compromised.

Higher demands are placed on the windows of housings for PTZ and dome cameras. Not only must the windows be specially shaped in the form of a bubble, but they also must have high clarity because imperfections such as dirt particles can be magnified, particularly when cameras with high zoom factors are installed. In addition, if the thickness of the window is uneven, a straight line may appear curved in the resulting image. A high-quality bubble should have very little impact on image quality, irrespective of the camera's zoom level and lens position.

Dome coverings or bubbles come in clear and smoked versions (Figure 15.8). Although smoked versions enable a more discreet

Figure 15.8 Examples of a clear and a smoked bubble for a PTZ dome camera. Although the smoked bubble makes it difficult to see in which direction the camera is positioned, it also reduces the amount of light and therefore the image quality.

installation, it is important to keep in mind that they also reduce the amount of light available to the camera and will therefore have an effect on the camera's light sensitivity. The darker shade of a smoked dome acts much like sunglasses do in reducing the amount of light that can pass through the covering.

15.3.3 Positioning of Fixed Camera

When installing a fixed camera in an enclosure, it is important to position the lens of the camera right up against the window to prevent any glare. Otherwise, reflections from the camera and the background will appear in the image. To reduce reflection, special coatings can be applied on any glass used in front of the lens (Figure 15.9).

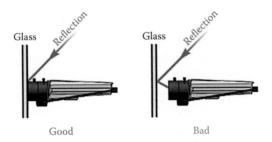

Figure 15.9 When installing a camera behind glass, correct positioning of the camera becomes important to avoid reflections.

15.3.4 Environmental Protection

The main environmental threats to a camera — particularly one that is installed outdoors — are cold, heat, water, and dust.

Housings with built-in heaters and fans (blowers) can be used in environments with low and high temperatures. In hot environments, cameras can be placed in enclosures that have active cooling with a separate heat exchanger.

To withstand water and dust, housings are carefully sealed. In situations where cameras might be exposed to acids, such as in the food industry, housings made of stainless steel are required. Some specialized housings can be pressurized, submersible, bullet-proof, or explosion-proof. Special enclosures also may be required for aesthetic considerations.

Other environmental elements include wind and traffic. To minimize vibrations, particularly on pole-mounted camera installations, the housing ideally should be small and securely mounted.

The terms "indoor housing" and "outdoor housing" often refer to the level of environmental protection. An indoor housing is used primarily to prevent the entry of dust and does not include a heater or fan. The terms are misleading because the location — whether indoors or outdoors — does not always correspond to the conditions at an installation site. A camera placed in an indoor freezer room, for example, will need an "outdoor housing" that has a heater.

15.3.5 Vandal and Tampering Protection

In some surveillance applications, cameras are at risk of hostile and violent attacks. Transportation (see Figure 15.10), schools, prisons, and retail environments are just some examples of areas where vandals or criminals may try to tamper with, redirect, destroy, spray paint, or remove cameras.

Although a camera or housing can never guarantee 100 percent protection from destructive behavior in every situation, there are a number of measures available that can help security managers deal with camera vandalism. The considerations to keep in mind include camera and housing design, mounting, placement, and intelligent video alarms.

15.3.5.1 The Goals of Vandal Protection

Many features and best practices can be implemented to increase protection against violence and vandalism. The important goals of vandal protection, regardless of actual technical implementation, include:

Figure 15.10 An example of a compact, fixed dome network camera specially designed for installations in mass transit vehicles such as buses and trains. It can withstand vibrations, dust, high humidity, and fluctuating temperatures.

- *Making it difficult.* Tampering with a video surveillance camera should be difficult, and perhaps even more importantly, it should be perceived as being difficult. Through camera design and placement, a vandal should be made to think twice before trying to interfere with a camera's operation.
- *Creating uncertainty.* If a vandal decides to attack a camera, he or she should be left with uncertainty as to whether the camera is actually destroyed or whether recording is still taking place.
- *Prolonging and delaying.* Even if it is not possible to protect a camera in the long run from a determined, hostile attack, it is worthwhile to make any attempts to redirect or destroy the camera very time consuming. Every second that passes increases the chance of a vandal's being discovered or simply giving up.
- *Detecting and sending alarm.* Intelligent functionality in a camera can detect that someone is tampering with its operation and notify operators. This allows operators to quickly alert staff in the field to deal with the problem by cleaning, adjusting, or replacing the camera or apprehending the vandal before the attack is completed.

15.3.5.2 Mechanical Design

Casings and related components made of metal provide better vandal protection than ones made of plastic. The shape of the housing or camera is another factor. A housing or a traditional fixed camera that protrudes from a wall or ceiling is more vulnerable to attacks (e.g., kicking or hitting) than more discreetly designed housings or casings for a dome camera. A dome's

Figure 15.11 Examples of fixed camera housings. Only the middle and right housings are classified as vandal resistant.

smooth, rounded covering makes it more difficult, for example, to block the camera's view by trying to hang a piece of clothing over the camera. The more a housing or camera blends into an environment or is disguised as something other than a camera — for example, an outdoor light — the better the protection against vandalism (Figures 15.11 through 5.13).

To provide additional vandal resistance in bubble coverings, they can be made of a durable, transparent material, such as polycarbonate plastic (the same material used to create bullet-proof glass). Increasing the thickness of a bubble improves the covering's ability to withstand heavy blows. However, the thicker a transparent covering is, the higher the chances of imperfections in clarity. This can translate to blurriness

Figure 15.12 An example of a fixed dome camera (left) and dome-shaped housing for fixed network cameras (right). Both are vandal resistant.

Figure 15.13 A typical vandal-resistant housing for a PTZ camera.

because imperfections, such as small particles in the glass, may be magnified at high zoom levels. Increased thickness also can create unwanted reflections and refraction of light, which will have a negative impact on image quality. Thicker coverings therefore should meet higher requirements if the effect on image quality is to be minimized.

Special coatings can also be applied to the bodies of cameras and housings to minimize the impact of graffiti. Housings with built-in wipers can help keep the transparent coverings clean.

15.3.5.3 Mounting

The way cameras and housings are mounted is also important. A traditional fixed network camera and a dome camera with a drop-ceiling mount are more vulnerable to attacks than a fixed or normal dome camera that is mounted flush to a ceiling or wall, where only the transparent part of the camera or housing is visible (Figure 15.14).

Screws that are not part of standard toolsets can make it more difficult for unauthorized people to dismount cameras and housings from walls and ceilings. The more unusual the screws, the better protection they provide. However, providing non-standard screws means that authorized staff need access to more rare and specific tools to mount, dismount, and move cameras around, which can be seen as inflexible by end users.

Another important consideration is how the cabling to a camera is mounted. Maximum protection is provided when the cable is pulled directly through the wall or ceiling behind the camera. In this way, there are no visible cables with which to tamper. If this is not possible, a metal conduit tube should be used to protect cables from attacks.

Figure 15.14 An example of a flush ceiling-mounted housing for fixed network cameras.

15.3.5.4 Camera Placement

Camera placement is also an important factor in deterring vandalism. Placing a camera out of reach on high walls or in the ceiling can prevent many spur-of-the-moment attacks. The downside may be the angle of view, which to some extent can be compensated by selecting a different lens.

15.3.5.5 Intelligent Video Protecting Cameras

Recent advances in intelligent video have made it possible to implement video analytics in network cameras and video management systems that help protect cameras against vandalism. Intelligent algorithms can detect if a camera has been redirected, obscured, or tampered with in other ways and can send alarms to operators in central control rooms or to staff in the field.

This is especially useful in installations of hundreds of cameras in demanding environments where keeping track of the proper functioning of all cameras is difficult. In situations where no live viewing takes place, intelligent video simplifies automatic surveillance by notifying operators when cameras have been tampered with. For more on intelligent video applications, see Chapter 14.

15.3.6 Types of Mounting

Cameras are necessary in all kinds of locations, and this requires a large number of variations in the type of mounting. A camera should be placed on a stable support to minimize camera movement. Because PTZ cameras move around, the action can cause image interference if the camera mounting is not secured properly. In outdoor situations, sturdy mounting equipment should always be used to avoid vibrations caused by strong winds.

15.3.6.1 Ceiling Mounts

Ceiling mounts are used primarily in indoor installations. The enclosure itself can be a:

- *Drop mount:* mounted directly at the surface of a ceiling and therefore completely visible.
- *Flush mount:* mounted inside the ceiling with only parts of a camera and housing (usually the bubble) visible.
- *Pendant mount:* hung from a ceiling like a pendant.

Figure 15.15 provides examples of each mounting type.

Figure 15.15 An example of a drop mount (left), a flush mount (middle), and a pendant mount (right).

Figure 15.16 An example of a corner wall mount.

15.3.6.2 Wall Mounts

Wall mounts are often used to mount cameras inside or outside a building. The housing is connected to an arm, which is mounted on a wall. Advanced mounts have an inside cable gland to protect the cable. Figure 15.16 depicts a corner wall mount.

15.3.6.3 Pole Mount

A pole mount (Figure 15.17) often is used together with a PTZ camera in locations such as a parking lot. This type of mount usually takes into consideration the impact of wind. The dimensions of the pole and the mount itself must be designed such that vibrations are minimized. Cables often are enclosed inside the pole, and outlets must be sealed properly. More advanced dome cameras have built-in electronic image stabilization to limit the effect of wind and vibrations.

15.3.6.4 Parapet Mount

Parapet mounts (Figure 15.18) are used for roof-mounted housings or to raise the camera for a better angle of view.

An IP rating consists of the letters IP followed by two digits. The first digit indicates the level of protection that an enclosure provides against access by solid foreign objects (e.g., a finger, a tool, or dust). The higher the number, the better the protection. The second digit relates to the level of protection against the access of liquids. Again, the higher the number, the better the protection. For example, an IP66 rating means it is dust tight and protects against access of heavy water jets. The minimum rating required for outdoor use is IP44. Most outdoor installations specify an IP66 rating. When there is no protection rating given with regard to either solid objects or liquids, a letter X is used, for example, IP2X. See Table 15.4.

Table 15.4 IP Ratings

First Digit	Description	Definition
Protection against Solid Foreign Objects		
0	Non-protected	No special protection.
1	Protected against solid objects greater than 50 mm	A large surface of the body such as the hand (no protection against deliberate access); solid objects exceeding 50-mm diameter.
2	Protected against solid objects greater than 12 mm	Fingers or other objects not exceeding 80 mm in length; solid objects exceeding 12-mm diameter.
3	Protected against solid objects greater than 2.5 mm	Tools, wires, etc. of diameter or thickness greater than 2.5 mm; solid objects exceeding 2.5-mm diameter.
4	Protected against solid objects greater than 1.0 mm	Wires or strips of thickness greater than 1.0 mm. Solid objects exceeding 1.0 mm.
5	Dust protected	Ingress of dust is not totally prevented, but dust does not enter in sufficient quantity to interfere with satisfactory operation of the equipment.
6	Dust-tight	No ingress of dust.

Second Digit	Description	Definition
Protection against Liquids		
0	Non-protected	No special protection.
1	Protected against dripping water	Dripping water (vertically falling drops).
2	Protected against dripping water when tilted up to 15°	Vertically dripping water shall have no harmful effect when the enclosure is tilted at any angle up to 15° from its normal position.

Continued.

Table 15.4 IP Ratings (Continued)

Second Digit	Description	Definition
3	Protected against spraying water	Water falling as spray at an angle up to 60° from the vertical shall have no harmful effect.
4	Protected against splashing water	Water splashed against the enclosure from any direction shall have no harmful effect.
5	Protected against water jets	Water projected from a nozzle against the enclosure from any direction shall have no harmful effect.
6	Protected against heavy seas	Water from heavy seas or water projected in powerful jets shall not enter the enclosure in harmful quantities.
7	Protected against the effects of immersion	Ingress of water in a harmful quantity shall not be possible when the enclosure is immersed in water under defined conditions of pressure and time.
8	Protected against submersion	The equipment is suitable for continuous submersion in water under conditions which shall be specified by the manufacturer.

15.3.7.2 NEMA Ratings

The National Electrical Manufacturers Association (NEMA) is a U.S.-based association that provides standards for electrical equipment enclosures. A NEMA rating specifies protection for specific environmental conditions (see Table 15.5).

15.3.7.3 IK Ratings

An IK rating specifies the degree of protection a housing provides against external mechanical impacts and is set according to CENELEC EN 50102 of June 1995 by the European Committee for Electrotechnical Standardization (Table 15.6).

15.3.7.4 IECEx and ATEX Certifications

When a camera will be installed in a potentially explosive environment, both the camera and the housing must comply with IECEx, which applies worldwide, or ATEX, which applies only in Europe. In such a situation, it is the environment that must be protected from any sparks or any unwanted effects coming from the camera to prevent or minimize the risk of an explosion. IECEx- or ATEX-certified equipment can be used in areas

Table 15.5 NEMA Ratings

NEMA Rating	Approx. IP Equivalent	Condition
In Non-Hazardous Locations:		
NEMA 1	IP10	Enclosures constructed for indoor use to provide a degree of protection to personnel against access to hazardous parts and to provide a degree of protection of the equipment inside the enclosure against ingress of solid foreign objects (falling dirt).
NEMA 2	IP11	Enclosures constructed for indoor use to provide a degree of protection to personnel against access to hazardous parts; to provide a degree of protection of the equipment inside the enclosure against ingress of solid foreign objects (falling dirt); and to provide a degree of protection with respect to harmful effects on the equipment due to the ingress of water (dripping and light splashing).
NEMA 3	IP54	Enclosures constructed for either indoor or outdoor use to provide a degree of protection to personnel against access to hazardous parts; to provide a degree of protection of the equipment inside the enclosure against ingress of solid foreign objects (falling dirt and windblown dust); to provide a degree of protection with respect to harmful effects on the equipment due to the ingress of water (rain, sleet, snow); and that will be undamaged by the external formation of ice on the enclosure.
NEMA 3R	IP14	Enclosures constructed for either indoor or outdoor use to provide a degree of protection to personnel against access to hazardous parts; to provide a degree of protection of the equipment inside the enclosure against ingress of solid foreign objects (falling dirt); to provide a degree of protection with respect to harmful effects on the equipment due to the ingress of water (rain, sleet, snow); and that will be undamaged by the external formation of ice on the enclosure.
NEMA 3S	IP54	Enclosures constructed for either indoor or outdoor use to provide a degree of protection to personnel against access to hazardous parts; to provide a degree of protection of the equipment inside the enclosure against ingress of solid foreign objects (falling dirt and windblown dust); to provide a degree of protection with respect to harmful effects on the equipment due to the ingress of water (rain, sleet, snow); and for which the external mechanism(s) remain operable when ice laden.

Continued

Table 15.5 NEMA Ratings (Continued)

NEMA Rating	Approx. IP Equivalent	Condition
NEMA 3X		Enclosures constructed for either indoor or outdoor use to provide a degree of protection to personnel against access to hazardous parts; to provide a degree of protection of the equipment inside the enclosure against ingress of solid foreign objects (falling dirt and windblown dust); to provide a degree of protection with respect to harmful effects on the equipment due to the ingress of water (rain, sleet, snow); that provide an additional level of protection against corrosion; and that will be undamaged by the external formation of ice on the enclosure.
NEMA 3RX		Fulfills the requirements of NEMA 3R and NEMA 3X.
NEMA 3SX		Fulfills the requirements of NEMA 3S and NEMA 3X.
NEMA 4	IP56	Enclosures constructed for either indoor or outdoor use to provide a degree of protection to personnel against access to hazardous parts; to provide a degree of protection of the equipment inside the enclosure against ingress of solid foreign objects (falling dirt and windblown dust); to provide a degree of protection with respect to harmful effects on the equipment due to the ingress of water (rain, sleet, snow, splashing water, and hose-directed water); and that will be undamaged by the external formation of ice on the enclosure.
NEMA 4X	IP56	Enclosures constructed for either indoor or outdoor use to provide a degree of protection to personnel against access to hazardous parts; to provide a degree of protection of the equipment inside the enclosure against ingress of solid foreign objects (windblown dust); to provide a degree of protection with respect to harmful effects on the equipment due to the ingress of water (rain, sleet, snow, splashing water, and hose-directed water); that provide an additional level of protection against corrosion; and that will be undamaged by the external formation of ice on the enclosure.
NEMA 5	IP52	Enclosures constructed for indoor use to provide a degree of protection to personnel against access to hazardous parts; to provide a degree of protection of the equipment inside the enclosure against ingress of solid foreign objects (falling dirt and settling airborne dust, lint, fibers, and flyings); and to provide a degree of protection with respect to harmful effects on the equipment due to the ingress of water (dripping and light splashing).

Table 15.5 NEMA Ratings (Continued)

NEMA Rating	Approx. IP Equivalent	Condition
NEMA 6	IP67	Enclosures constructed for either indoor or outdoor use to provide a degree of protection to personnel against access to hazardous parts; to provide a degree of protection of the equipment inside the enclosure against ingress of solid foreign objects (falling dirt); to provide a degree of protection with respect to harmful effects on the equipment due to the ingress of water (hose-directed water and the entry of water during occasional temporary submersion at a limited depth); and that will be undamaged by the external formation of ice on the enclosure.
NEMA 6P	IP67	Enclosures constructed for either indoor or outdoor use to provide a degree of protection to personnel against access to hazardous parts; to provide a degree of protection of the equipment inside the enclosure against ingress of solid foreign objects (falling dirt); to provide a degree of protection with respect to harmful effects on the equipment due to the ingress of water (hose-directed water and the entry of water during prolonged submersion at a limited depth); that provide an additional level of protection against corrosion; and that will be undamaged by the external formation of ice on the enclosure.
NEMA 12	IP52	Enclosures constructed (without knockouts) for indoor use to provide a degree of protection to personnel against access to hazardous parts; to provide a degree of protection of the equipment inside the enclosure against ingress of solid foreign objects (falling dirt and circulating dust, lint, fibers, and flyings); and to provide a degree of protection with respect to harmful effects on the equipment due to the ingress of water (dripping and light splashing).
NEMA 12K	IP52	Enclosures constructed (with knockouts) for indoor use to provide a degree of protection to personnel against access to hazardous parts; to provide a degree of protection of the equipment inside the enclosure against ingress of solid foreign objects (falling dirt and circulating dust, lint, fibers, and flyings); and to provide a degree of protection with respect to harmful effects on the equipment due to the ingress of water (dripping and light splashing).

Continued

Table 15.5 NEMA Ratings (Continued)

NEMA Rating	Approx. IP Equivalent	Condition
NEMA 13	IP54	Enclosures constructed for indoor use to provide a degree of protection to personnel against access to hazardous parts; to provide a degree of protection of the equipment inside the enclosure against ingress of solid foreign objects (falling dirt and circulating dust, lint, fibers, and flyings); to provide a degree of protection with respect to harmful effects on the equipment due to the ingress of water (dripping and light splashing); and to provide a degree of protection against the spraying, splashing, and seepage of oil and non-corrosive coolants.
In Hazardous Locations:		
NEMA 7		Enclosures constructed for indoor use in hazardous (classified) locations classified as Class I, Division 1, Groups A, B, C, or D as defined in NFPA 70.
NEMA 8		Enclosures constructed for either indoor or outdoor use in hazardous (classified) locations classified as Class I, Division 1, Groups A, B, C, and D, as defined in NFPA 70.
NEMA 9		Enclosures constructed for indoor use in hazardous (classified) locations classified as Class II, Division 1, Groups E, F, or G, as defined in NFPA 70.
NEMA 10		Enclosures constructed to meet the requirements of the Mine Safety and Health Administration, 30 CFR, Part 18.

Table 15.6 IK Ratings

IK Rating for Protection against the Factors Listed Below	IK01	IK02	IK03	IK04	IK05	IK06	IK07	IK08	IK09	IK10
Energy at the impact (Joule)	0.14	0.2	0.35	0.5	0.7	1	2	5	10	20
Mass (Kg)	0.25	0.25	0.25	0.25	0.25	0.25	0.5	1.7	5	5
Stroke down (mm)	56	80	140	200	280	400	400	300	200	400

Figure 15.20 Example dome camera and housing unit designed and certified for use in potentially explosive atmospheres and harsh environmental conditions. The camera housing is made of stainless steel and thus is ideally suited for use in offshore and onshore environments. Stainless steel screws and mounting bracket are incorporated, ensuring a totally corrosion-free unit. (Courtesy of Dexter, Dutch Explosafety Center.)

Figure 15.21 ATEX and CE logos.

where flammable liquids, vapors, gases, or combustible dusts are likely to occur in quantities sufficient to cause a fire or explosion (Figure 15.20). Examples of such areas include gas stations, oil platforms and refineries, chemical processing plants, printing industries, gas pipelines and distribution centers, grain handling and storage, aircraft refueling and hangars, and hospital operating theaters.

IECEx is part of the International Electrotechnical Commission, whereas ATEX, which stands for Atmospheres Explosibles in French, stems from the European Union's Directive 94/9/EC. ATEX-certified housings bear the "Ex" hexagon logo and CE mark (see Figure 15.21).

15.4 Storage and Server Considerations

Setting up the server and storage part of a network video system can be anything from a trivial five-minute task to a very complex and time-

Figure 15.22 A small system.

consuming activity, depending on the system size and requirements. When designing the server and storage systems, there are a few basic decisions that must be made — most importantly, the performance of the recording servers and whether a central or distributed architecture will be used. In addition, the amount of storage must be calculated. The performance of the server required, potential system architectures, and storage system also depend on the video management software used. For more information on storage and servers, see Chapter 11.

15.4.1 Small System: 1 to 30 Cameras

A small system usually consists of one server running a surveillance application that records the video to a local hard disk. The video is viewed and managed by the same server. Although most viewing and management will be done at the server, a client (local or remote) could be connected for the same purpose (Figure 15.22).

15.4.2 Medium System: 25 to 100 Cameras

A typical, medium installation has a server with additional storage attached to it (Figure 15.23). The storage used is usually configured with RAID to increase performance and reliability. The video is normally viewed and managed from a client rather than from the recording server itself.

15.4.3 Large Centralized System: 50 to 1000+ Cameras

A large-sized installation requires high performance and reliability to manage the large amount of data and bandwidth. This requires multiple servers with dedicated tasks (Figure 15.24). A master server controls the

Figure 15.23 A medium system.

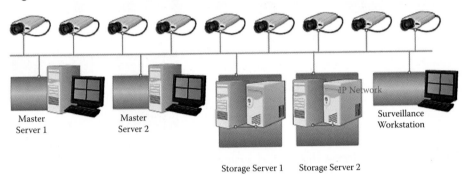

Figure 15.24 A large centralized system.

system and decides what kind of video is stored at what storage server. Because there are dedicated storage servers, it is possible to do load balancing. In such a setup, it is also possible to scale up the system by adding more storage servers when needed and also do maintenance without bringing down the whole system.

15.4.4 Large Distributed System: 25 to 1000+ Cameras

When multiple sites require surveillance with centralized management, distributed recording systems can be used (Figure 15.25). Each site records and stores the video from local cameras. The master controller can view and manage recordings at each site.

15.4.5 Dimensioning the Server

A PC server used for video management should be properly dimensioned. Each server in a video surveillance system can handle a certain number of cameras and a certain number of frames per second, depending on the CPU (central processing unit), network card, and internal RAM (random

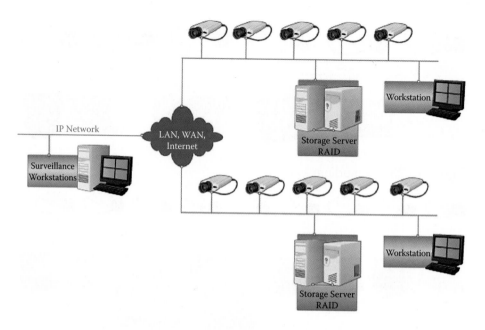

Figure 15.25 A large distributed system.

access memory). Table 15.7 gives an indication of what a typical video surveillance system requires.

15.4.6 Calculating Storage

Calculating the appropriate amount of storage is a very important task when designing a video surveillance system. However, it is not an exact science because the size of video files depends on the complexity and amount of motion in a scene. Some guidelines, along with an example of how the amount of storage can be limited by changing some parameters in the system, are provided below.

15.4.6.1 Calculating Storage Needs

A proper calculation of the storage required in a network video system is essential for the success of the network video project. Factors to consider when calculating storage needs include:

- Number of cameras
- Number of hours per day the camera will be recording
- How long the data must be stored

Table 15.7 Some Installation Scenarios with Recommended Server Hardware

Target Cameras	FPS	Hard Disks	Required Server Bandwidth (Mbit)	CPU (GHz)	RAM
5	5	1	100	2	512
	10	1	100	2.5	512
	20	1	100	3	512
10	5	2	100	2.5	512
	10	2	100	3	512
	20	2	100	3.4	768
15	5	2	100	3	512
	10	3	1000	3.4	768
	20	3	1000	Xeon dual 3.0	768
20	5	3	100	3.4	768
	10	4	1000	Xeon dual 2.8	768
	20	4	1000	Xeon dual 3.0	1024
25	5	3	100	Xeon dual 2.8	768
	10	4	1000	Xeon dual 3.0	1024
	20	4	1000	Xeon dual 3.4	1024

Note: All calculations are based on 640×480 image size at 25 percent compression using Motion JPEG. The CPU recommendation is based on Pentium4/Xeon dual processors. Please note that the requirements can vary with different video management software applications.

- Motion detection (event) only or continuous recording
- Other parameters such as frame rate, compression, image quality, and complexity

Note that the calculation examples below are examples only and do not take into consideration any overhead or other technical issues that may result in a higher file size than mentioned.

The calculation examples do not consider storage space for the operating system or video management software.

JPEG. For JPEG which consists of individual files, storage requirements vary by changing the frame rate, resolution, and compression. Cameras 1, 2, and 3 in Table 15.8 have different storage requirements according to their frames per second (fps) and resolution settings.

Table 15.8   Example of Storage Requirements Calculation for JPEG

Camera	Resolution	Image Size (KB)	Frames per Second	MB per Hour	Hours of Operation	GB per Day
No. 1	CIF	13	5	234	8	1.9
No. 2	CIF	13	15	702	8	5.6
No. 3	4CIF	40	15	2160	12	26

Table 15.9   Example of Storage Requirements Calculation for MPEG-4

Camera	Resolution	Bit Rate (Kbps)	Frames per Second	MB per Hour	Hours of Operation	GB per Day
No. 1	CIF	170	5	76.5	8	0.6
No. 2	CIF	400	15	180	8	1.4
No. 3	4CIF	880	15	396	12	5

Note: In this example, MPEG-4 provides a 75 percent savings in network bandwidth and storage, compared with using JPEG, as in the example above.

Calculation:

Image size × Frames per second × 3600 s = KB per hour / 1000 = MB per hour

MB per hour × Hours of operation per day / 1000 = GB per day

GB per day × Requested period of storage = Storage need

For the data given in Table 15.8, the total requirement for the three cameras and 30 days of storage is 1,002 GB.

MPEG-4. In MPEG-4, the images are part of a continuous data stream, that is, not individual files. It is the bit rate (8 bits = 1 byte), which measures the amount of video data transmitted, that determines the corresponding storage requirements. The bit rate is a result of specific frame rate, resolution, and compression, as well as the level of motion in the scene (see example in Table 15.9).

Calculation:

Bit rate/8 × 3600 s = KB per hour / 1000 = MB per hour

MB per hour × Hours of operation per day / 1000 = GB per day

GB per day × Requested period of storage = Storage need

For the example in Table 15.9, the total requirement for the three cameras and 30 days of storage is 205 GB.

15.4.6.2 An Example

There are many factors that have great impact on how much storage is needed. The factors can be broken down into four groups:

1. Video quality: frames per second, resolution, etc.
2. Video compression type: Motion JPEG, MPEG-4, H.264
3. Scenery: image complexity (e.g., gray wall or a forest), lighting conditions, and amount of motion (office environment or crowded train stations)
4. System size and requirements: number of cameras, how long the video is stored, and if recordings are continuously recorded or only triggered by events

To estimate the amount of storage required, all the above parameters must be analyzed and defined. Simple adjustments to the above factors can yield considerable savings in a storage system. Consider the case of using a camera with the following parameters: high video quality (30 fps, 640×480 resolution, and low-compression MPEG-4), continuous recording, and video retention for seven days. This would require approximately 70 GB. Adjusting a couple of parameters would drastically reduce the storage needs.

Consider the following changes:

- Continuous recording during office hours, and triggered recording during nighttime → 25 GB
- Reducing frame rate to 12 fps → 10 GB
- Increasing compression (reducing quality) → 6 GB
- Recording only events triggered by video motion detection (calculated 10 percent motion) → 1 GB
- Reducing frame rate to 6 fps → 550 MB
- Storing only for 2 days → 150 MB

Using all the above adjustments will most likely be sufficient to identify any persons and incidents in, for example, a reception area. Furthermore, it also will be much faster and easier to locate an incident in 150 MB of data than in 70 GB of data.

15.5 Designing the Network Bandwidth

When designing a network video system, it is important to properly design the network and associated bandwidth. Although video system network design was a big challenge only a few years ago, today's gigabit networks make it less challenging because, if designed correctly, they can easily cope with today's large amounts of network video. Network video products utilize network bandwidth based on their configuration. Bandwidth usage, as with storage, depends on image resolution, compression, and frame rate, as well as the complexity of the scene.

15.5.1 Limiting the Bandwidth

Today's gigabit networks can easily cope with the demands of network video systems. There are also several technologies available that enable the management of bandwidth consumption, including:

- *Switched networks and VLANs.* Using virtual local networks (VLANs) on a switched network — a common networking technique today, the same physical computer and video surveillance network can be separated into two autonomous networks. Although these networks remain physically connected, the network switch logically divides them into two virtual and independent networks.
- *Quality of service (QoS).* Using QoS will guarantee a certain bandwidth to a specific application or to a certain camera in a video surveillance system.
- *Event-driven frame rate.* A rate of up to 25 or 30 fps on all cameras at all times is above the level required for many applications. With the configuration capabilities and built-in intelligence of the network camera/video encoder, frame rates under normal conditions can be set lower — for example, five frames per second — which will dramatically decrease bandwidth consumption. In the event of an alarm (e.g., if motion detection is triggered), the recording frame rate can be increased automatically. In many cases, it is sufficient to have the camera send video over a network only if the video is worth recording; the rest of the time nothing needs to be sent.

15.5.2 Network Latency

In a video surveillance system, it may be important to consider latency. This is especially true when the video is being monitored live and when a

PTZ camera is being operated. A well-designed network should have very little latency, typically in the hundred-millisecond range. In a network with many hops, latency can be upwards of a second or higher, which can present a problem.

15.6 System Design Tools

Designing a network video system requires making a lot of choices and fine-tuning various settings. Attached to this book is a DVD of a system design tool (Figure 15.26). The tool includes functionalities for designing a video surveillance project that consists of many cameras. By selecting the frame rates, resolution, compression, and image scenario, the bandwidth and storage requirements can be calculated. Individual projects can be created and stored.

In the tool, there are also advanced features for helping users understand the effect of certain image scenarios and what a certain resolution and frame rate mean, using recorded video as examples (Figure 15.27).

Figure 15.26 The system design tool includes advanced project management functionality that enables calculation of bandwidth and storage requirements for a large and complex system.

Figure 15.27 The system design tool provides help in understanding what effect frame rate, resolution, and compression have on the quality of the video, as well as the resulting bandwidth.

15.7 Legal Aspects

Video Surveillance
in Operation

Video surveillance can be restricted or prohibited by laws that vary from country to country. It is advisable to check the laws in the local region before installing a video surveillance system.

There may be legislation or guidelines covering the following:

354

- *License.* It may be necessary to register or get a license from an authority to conduct video surveillance, particularly in public areas.
- *Purpose of the surveillance equipment.* Is it in accordance with what is permitted by the laws in the area?
- *Position or location of the equipment.* Is it positioned or located in such a way that it only monitors the spaces the equipment is intended to cover? If unintended areas are covered, consultations with the owners of such spaces may be required. There may be rules covering areas where video surveillance is prohibited, for example, restrooms and changing rooms in a retail environment.
- *Notification.* Signs may have to be placed to warn the public that they are entering a zone covered by surveillance equipment, and there may be rules regarding the signage.
- *Quality of images.* There may be rules regarding the quality of images, which can affect what may be permitted or acceptable for use as evidence in court.
- *Video format.* Police authorities may require that the video format be one that they can handle.
- *Information provided in recorded video.* Video recordings, for example, may be required to have time and date stamps.
- *Processing of images.* There may be rules regulating how long images should be retained, who can view such images, and where recorded images can be viewed.
- *Drawings.* There may be requirements for drawings of where cameras are placed.
- *Personnel training.* There may be regulations that require operator training in security and disclosure policies as well as privacy issues.
- *Access to and disclosure of images to third parties.* There may be restrictions on who can access the images and how images can be shown. For example, if video will be disclosed to the media, images of individuals may have to be disguised or blurred.
- *Monitoring and recording of audio.* A permit may be required for recording audio in addition to video.
- *Regular system checks.* There may be guidelines on how often and thoroughly a company should perform system checks to make sure all equipment is operating as it should.
- *Audit trail.* This means having the ability to show who used the system, at which time, and for what purpose. In addition, proof of a video's authenticity may be required by methods such as watermarking.

Chapter 16

The Cost of a Network Video System

Before installing a video surveillance system, end users must decide between an analog or an IP-based system. Both systems have technical advantages and disadvantages. However, installers and security experts know perfectly well that end users often base their final decision on price rather than technical merit.

Advocates of analog video surveillance usually argue that network cameras are about 50 percent more expensive than analog ones. However, this is only half the truth, and it does not reflect the fact that cameras are only one component of the overall video surveillance system. Buyers also should factor in costs for system maintenance, video recording and playback, cameras, installation, configuration, training, and cable infrastructure.

Many installers, integrators, and security consultants who work with IP-based video systems are convinced that the overall costs to install IP-based video systems are lower than with analog. In the spring of 2007, an independent researcher conducted a study aimed at bringing some clarity to this area by comparing the two types of video surveillance systems:

1. *Analog surveillance system:* analog cameras and DVR-based recording
2. *IP-based video surveillance system:* network cameras, IP infrastructure, server, software, and storage

The objective of the study was to develop an understanding of the total cost of ownership in a typical video surveillance scenario for the two types of systems. A structured research methodology was developed and validated through interviews and through a standard bid request process with security integrators and value-added resellers. This chapter describes the procedures and findings of the research.

16.1 Research Approach

The primary objective of the study was to develop an unbiased understanding of the total cost of ownership (TCO) for two types of video surveillance systems: (1) an analog surveillance system (analog cameras and DVR based recording), and (2) a fully digital IP-based surveillance system (network cameras, IP infrastructure, server, video management software, and storage).

To make the study as impartial and as balanced as possible, a structured research approach was developed. It included a step-by-step validation of each project phase by security integrators, value-added resellers, and industry analysts. Definition of the cost components, deployment scenarios, and assumptions were developed with, and scrutinized by, these study participants with the objective of making the research approach and study results as fair and as unbiased as possible. In addition to interviews, an industry-standardized approach was used to collect cost data, which included the development of an RFP (Request For Proposal), that is, what an end user would likely provide to a security integrator to solicit a system proposal or project bid that would contain itemized costs components, and then solicitation of responses or "project bids" to collect structured cost data.

The research approach was divided into three phases:

1. Develop, validate, and refine the typical video surveillance scenario and cost comparison framework with research participants
2. Use structured interviews and standard methods (e.g., an RFP and bids) to collect quantitative cost data
3. Review, validate, and synthesize the findings

Nonquantifiable observations and cost considerations that differentiated the two types of surveillance systems were not included in the total cost comparison but are detailed at the end of this chapter.

A dozen interviewees from different geographic regions in North America participated in the study by providing input on study components, feedback, validation, and cost data (in the form of bid responses). Several preliminary interviews were conducted to develop and validate the total cost of ownership definition for the study. The emphasis was on quantifiable "hard costs" that could be supplied by study participants with a minimum of interpretation or ambiguity.

16.2 Typical System

Several scenarios were considered and discussed with study participants. They included a small office/indoor surveillance scenario (e.g., four to eight fixed cameras), a mid-sized "mainstream" scenario (e.g., fixed and PTZ cameras, both indoor and outdoor), as well as a "large site" scenario (up to several hundred cameras with multiple geographic locations).

Several observations were collected on the merits of each system, with a consensus settling on the mid-sized scenario, which was ideal for the study in both scope and complexity, as well as a scenario that offered no clear cost advantages for either an IP- or an analog-based surveillance system. For example, several study participants shared the observation that a "large site" scenario might have inherent cost advantages for an IP-based system. This is due to the possibility of utilizing a shared network infrastructure for various data types including control, video, and audio, as well as perceived advantages of an all-IP-network-based system for simplified remote "end-to-end" management down to each individual camera location.

The next step was to define a typical mid-sized video surveillance scenario that offered a sufficient "apples-to-apples" framework for comparing individual cost components for the two cost categories and surveillance systems that were previously defined. The selected video surveillance system was one for a small- to mid-sized school campus. The "school surveillance" scenario was defined and reviewed by study participants and refined into a set of system requirements, operational assumptions, and individual cost elements that could be developed into structured and unbiased interview material for the collection of cost data. It also was determined that providing study participants with a Request for Proposal (RFP) for this scenario would most efficiently facilitate the collection of unambiguous and impartial cost data.

To make the comparison as unbiased as possible, selecting the number of cameras for the installation should not be advantageous for either the analog system or the IP-based system. Because analog systems typically are based on multiples of 16 cameras — because a DVR has 16 inputs — it was important not to select 16, 32, or 48. At the same time, IP-based systems have an advantage at 17, 33, and 49. Therefore, 40 cameras were selected as a fair case for both types of systems.

In addition, the number was typical for a mid-sized system, and this was validated during the interviews. When collecting cost information from study participants, they were required only to meet the "customer requirements" for the school facility installation. No cameras, wiring, or infrastructure were said to exist on the premises; that is, all new data and power cabling would be required. The study participants had full autonomy in selecting the equipment and setting the pricing for configuration, service, and upgrades, among other things.

The highlights of the customer requirements as outlined in the RFP are as follows:

- Facility:
  - Single-building school
  - Existing building
- Number of cameras:
  - 30 indoor fixed dome cameras
  - 5 outdoor fixed dome cameras
  - 5 outdoor pan/tilt/zoom (PTZ) cameras
  - All cameras need to be vandal-proof
- Recording:
  - 12 hours of recording a day
  - 4 frames per second (fps) continuous recording
  - 15 fps recording on alarm/video motion detection
  - CIF resolution (352 × 240)
  - Retention of video for 12 days
- Wiring:
  - No existing data, coax, or power wiring
  - Network switches (wiring closets) or multi-camera power supplies
  - Plenum airspace above all areas (for cabling, plenum wiring required)
  - Cat5e adequate for data wiring
  - PoE (Power over Ethernet) switches can be located in the storage area, allowing for less than 250 ft. PoE cabling to network cameras
- Monitoring location and equipment placement:
  - Main network hub and camera viewing (location of monitor, server/DVR) in area labeled Administration (representing several offices)

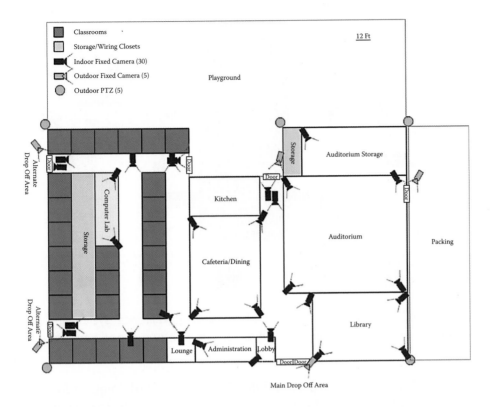

Figure 16.1 Illustration of the fictitious school building with proposed camera placements.

- Network switches (wiring closets) or multi-camera power supplies can be placed in any gray-shaded area on the diagram, as shown in Figure 16.1.
- Other:
  - No special illuminators required
  - No audio surveillance required

Figure 16.1 illustrates the school facility along with the camera placements.

Once a typical video surveillance scenario was defined, validated, and refined, the next steps were to develop and define a list of cost-contributing components.

16.3 Cost to Purchase and Install

The next step involved determining the specific elements that contribute to the total cost of putting an IP or analog surveillance system into service within the previously defined scenario (i.e., school facility).

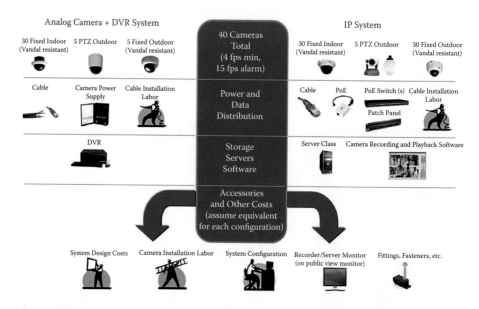

Figure 16.2 Various components of a video surveillance system include cameras, power and data distribution, storage software and servers, and installation.

To deal with this aspect and develop a structured "request for bid" that could be used to collect individual costs from study participants, a set of purchase and installation cost-contributing components was developed. First, costs that were considered equivalent for either the IP or analog system were determined and validated. Second, a basic grouping of cost components was defined and validated as in Figure 16.2.

When collecting cost information from study participants, they were required only to meet the "customer requirements" for the school facility installation. Otherwise, the study participants had full autonomy in selecting the equipment and setting — for example, the pricing for configuration, service, and upgrades. The cost information was supplied in the form of quotes or "bids."

16.4 Research Results

The resulting quotes from the participating systems integrators revealed some interesting information. The quoted costs included the cost to acquire and install the equipment, and using the average value of all quotes, the findings were as follows.

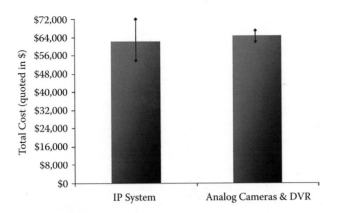

Figure 16.3 The total cost of the IP-based surveillance system was, on average, 3.4 percent lower than the analog system. The line laid over each column indicates the spread of the quotes, which was much larger for the IP-based system.

16.4.1 Overall Comparison

Looking at the total cost of the system and comparing the average cost for the IP system and the average cost for the analog system, it was interesting to find that the total cost for the IP system had a 3.4 percent lower total cost of ownership.

Looking further at the different bids revealed that the lowest-quoted IP system had a 25.4 percent lower cost than the lowest-priced analog/DVR system, and the highest-quoted IP system had an 11.5 percent higher cost than the highest-priced analog/DVR system. The findings are presented in Figure 16.3.

It is interesting to note the flexibility of the IP-based system as represented by the wide spread in the quotes. The reason is the wide flexibility that the IP technology provides, because wide-ranging possibilities and choices exist when it comes to using PoE, cabling, network, and server platforms. In an analog system, there is very little flexibility; hence, most quotes came in at close to the same price. That is quite typical for a mature market.

16.4.2 Detailed Cost Comparison

The split of the cost in the IP-based system showed that it is quite different from the analog/DVR system, as outlined in Figure 16.4.

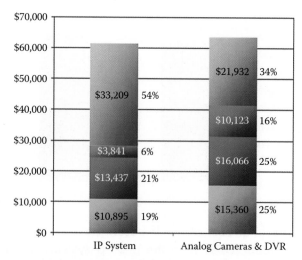

Figure 16.4 Looking at the details, it becomes obvious that the camera portion in the IP-based system is more expensive than the analog counterpart, but all other parts of the IP-based system provide considerable savings.

In comparing the detailed costs, the conclusions were that:

- The network cameras are 50 percent more expensive than their analog counterparts and can be half of the system cost in the IP-based system, whereas analog cameras make up only a third of the cost in the analog/DVR system.
- The cost of cabling is almost three times more expensive in the analog system than in the IP-based system. This is because an analog system requires separate cables for power and video signals. In addition, separate cabling must be installed to control analog PTZ cameras. In an IP system, the cables used for data transmission also can be used to control PTZ cameras and carry power at the same time.
- The cost of recording and monitoring is similar. The quality, available service, and maintenance contracts for a PC server in the IP system are often better than for a DVR.
- The installation, configuration, and training costs are almost 50 percent higher in the analog system than in the IP-based system.

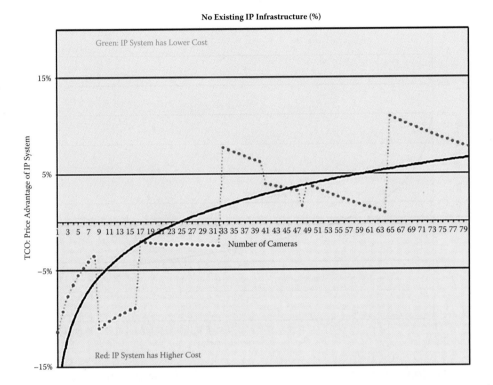

Figure 16.5 Graph showing the cost difference (in percentages) as a function of the number of cameras when there is no preexisting IP infrastructure to use.

16.4.3 Cost as a Function of the Number of Channels

A general consensus about IP-based systems is that the larger the system is, the more favorable the cost will be for an IP system than for an analog system. So where would the breakpoint be? That is, for what size system would IP be lower in cost than analog, and does the difference increase as the size of the system increases? Based on the research data and other information, the cost as a function of the number of cameras was calculated. Figure 16.5 provides the results.

The result shows that an IP-based system is lower in cost than an analog system when involving more than 32 cameras. The costs between the two systems are quite similar when involving between 16 and 32 cameras. In the case above, the assumption was that no cabling infrastructure was in place.

In many buildings today, an IP infrastructure already exists, which the surveillance system can piggyback on. What would be the costs in

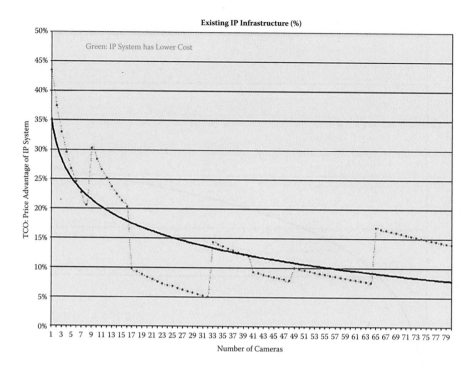

Figure 16.6 Graph showing the cost difference (as a percentage) as a function of number of cameras when an IP infrastructure exists. The results show that the cost of the IP-based system is always lower.

such a case? With the cabling and installation costs removed from the calculations, the results become those shown in Figure 16.6.

The results show that if an IP infrastructure is already in place, the IP-based system always has a lower cost than the analog system, even for very small systems.

16.4.4 Additional Observations

During the research portion of the study, several interview participants provided additional nonquantifiable observations and cost considerations that differentiated the two types of video surveillance systems. Although the considerations were not included in the total cost comparison, they were considered important baseline differences by the interview participants and are therefore detailed here:

- Scalability is superior in IP-based systems, where it is possible to add one camera at a time.

- Flexibility is greater in IP-based systems because moving a camera means only moving a network drop if PoE is used.
- The image quality of network cameras is superior to that of analog.
- Megapixel cameras are beginning to be specified, which can only be addressed by network cameras.
- The IP infrastructure is often already in place and can be used by a network video system.
- Analog coax cabling is much more difficult to troubleshoot than an IP network.
- System design costs are typically included at no additional cost.
- An IP system can be remotely serviced (e.g., adjusted or diagnosed over the network).
- Brand-name PC servers used in IP systems often have superior warranty and service plans compared with DVRs.
- The price of IT equipment is likely to drop faster than that of analog equipment.

16.5 Conclusions

Research conducted with security integrators, value-added resellers, and industry analysts yielded some major findings:

- IP-based systems of 40 cameras have a lower total cost of ownership than analog-based systems. Based on a typical deployment scenario as outlined in the study, the cost to acquire and install an IP-based system is, on average, 3.4 percent lower than an analog-based solution. Thirty-two cameras is the break-even point for IP-based systems versus analog systems. Based on common scenarios for costs, an IP-based surveillance system is lower in cost than an analog system if more than 32 cameras are involved. For 16 to 32 cameras, the costs are quite similar and may even be slightly lower for analog systems.
- If an IP infrastructure is already in place, the IP system always has the lower cost, no matter the size of the system.
- Many nonquantifiable advantages for IP systems exist, including improved image quality, better maintenance and service, and increased flexibility. In addition, the price of IT equipment is expected to fall faster than that of analog CCTV equipment, making the comparison even more favorable in the future for IP-based surveillance systems.

Index

X

Z